T0146172

Analysis of Global Management of Air Force War Reserve Materiel to Support Operations in Contested and Degraded Environments

KRISTIN F. LYNCH, ANTHONY DeCICCO, BART E. BENNETT,
JOHN G. DREW, AMANDA KADLEC, VIKRAM KILAMBI, KURT KLEIN,
JAMES A. LEFTWICH, MIRIAM E. MARLIER, RONALD G. MCGARVEY,
PATRICK MILLS, THEO MILONOPOULOS, ROBERT S. TRIPP,
ANNA JEAN WIRTH

Prepared for the Department of the Air Force
Approved for public release; distribution unlimited

PROJECT AIR FORCE

For more information on this publication, visit www.rand.org/t/RR3081

Library of Congress Cataloging-in-Publication Data is available for this publication.
ISBN: 978-1-9774-0377-3

Published by the RAND Corporation, Santa Monica, Calif.
© 2021 RAND Corporation
RAND® is a registered trademark.

Cover photo credit: Senior Airman Jeremy McGuffin/U.S. Air Force

Support RAND
Make a tax-deductible charitable contribution at
www.rand.org/giving/contribute

www.rand.org

Preface

Adversaries have developed capabilities that may restrict or deny the U.S. military's access to given areas, creating an operational environment that is very different from the environment of the past 30 years. Prepositioning war reserve materiel (WRM) may help mitigate vulnerabilities associated with operating in a contested, degraded, or operationally limited environment, and the WRM may serve as a signal of U.S. presence that helps deter aggression and coercion by adversaries.

The U.S. Air Force has centralized some aspects of managing a limited amount of WRM under the 635th Supply Chain Operations Wing.[1] For those limited capabilities, the Air Force has recognized some improvement in support. The objective of the analysis described in this report is to evaluate management approaches and global prepositioning strategies for WRM postures for areas of responsibility in contested, degraded, or operationally limited environments. We consider classes of WRM beyond those currently managed by the 635th Supply Chain Operations Wing; however, we do not include munitions. In this report, we describe conditions under which global management practices are advantageous and then propose methods that a global manager of WRM could employ to improve support of air component operational warfighting demands. Specifically, we suggest ways to standardize and validate determination processes for WRM requirements, establish a WRM capability prioritization schema, relate WRM priority to positioning postures, analyze trade-offs through modeling, and assess partner-nation risk. This research should be of interest to personnel involved in logistics, sustainment, or operational planning in the U.S. Air Force.

The research reported here was commissioned by the U.S. Air Force and conducted within the Resource Management Program of RAND Project AIR FORCE as part of a fiscal year 2018 project titled "Analysis of Global War Reserve Materiel Postures to Support Contested and Degraded Combat Operations Across Areas of Responsibility."

RAND Project AIR FORCE

RAND Project AIR FORCE (PAF), a division of the RAND Corporation, is the Department of the Air Force's (DAF's) federally funded research and development center for studies and analyses. PAF provides the DAF with independent analyses of policy alternatives affecting the development, employment, combat readiness, and support of current and future air, space, and cyber forces. Research is conducted in four programs: Strategy and Doctrine; Force

---

[1] This work was completed before the establishment of the U.S. Space Force. References to the Air Force throughout this document refer to the U.S. Air Force.

Modernization and Employment; Manpower, Personnel, and Training; and Resource Management. The research reported here was prepared under contract FA7014-16-D-1000.

Additional information about PAF is available on our website: www.rand.org/paf

This report documents work originally shared with the U.S. Air Force on October 22, 2018. The draft report, issued on September 28, 2018, was reviewed by formal peer reviewers and DAF subject-matter experts.

Contents

Figures

Tables

Summary

Adversaries have developed capabilities that may restrict or deny the U.S. military's access to given areas, creating an operational environment that is very different from the environment of the past 30 years. Prepositioning war reserve materiel (WRM) may help mitigate vulnerabilities associated with operating in a contested, degraded, or operationally limited (CDO) environment, and the WRM may serve as a signal of U.S. presence that helps deter aggression and coercion by adversaries.

The objective of this analysis is to evaluate management approaches and global prepositioning strategies for WRM postures for areas of responsibility in CDO environments. There are different ways to manage WRM in a resource-constrained environment, and each has associated risk. In this analysis, we focus on global management—a strategy in which processes are controlled centrally, with decentralized execution for the core functions. For WRM, core functions include acquisition, storage and maintenance, distribution, and assessments.

In this report, we describe conditions under which global management practices are advantageous and then propose methods that a global manager of WRM could employ to improve support of air component operational warfighting demands. Specifically, we suggest ways to standardize and validate determination processes for WRM requirements, establish a WRM capability prioritization schema, relate WRM priority to positioning postures, analyze trade-offs through modeling, and assess partner-nation risk. We began our effort by reviewing prior analyses of WRM and CDO environments. We then conducted an academic literature review of global management and investigated the use of global management in industry. We reviewed other services' WRM practices and then compared current U.S. Air Force WRM practices with how WRM could be globally managed. Finally, in this report, we suggest processes, analytic capabilities, tools, and systems that the Air Force could employ to improve global management of WRM. Our analytic approach is shown in Figure S.1.

Figure S.1. Analytic Approach

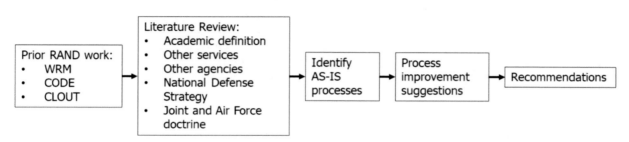

NOTE: CLOUT = coupling logistics with operations to meet uncertainty and the threat; CODE = combat operations in denied environments.

From our review of academic literature, industry practices, and other services processes, we gleaned several overarching themes related to management practices (see Table S.1). For example, a decentralized management strategy facilitates a rapid response to changes in demand but can generate information distortions,[1] leads to cost inefficiencies, and promotes organizational fragmentation. On the other hand, a more centralized management strategy promotes information-sharing; process standardization; and operational flexibility to adapt to major disruptions, such as political restrictions, natural disasters, or terrorist attacks, at one node by reallocating resources to other nodes within the supply network.[2] However, this flexibility comes at an expense. Centralized management practices may be slower to detect and respond to changes in operations, and products may not be tailored to local preferences. Centralized management practices also stifle motivation and inhibit the creativity that promotes innovation and problem-solving among frontline professionals, which can be especially limiting in high technology and service industries.[3] Generally, organizations have a mix of management strategies—some centralized and some decentralized, with each tailored to support the activity that the organization serves.

Table S.1. Common Themes in the Global Management Literature

When it is important to have . . .	It is usually advantageous to veer toward . . .	Because that solution enables and stimulates . . .
Responsiveness	Decentralization	Immediacy
Reliability	Centralization	Compliance
Efficiency	Centralization	Syndication

SOURCE: Adapted from Vantrappen and Wirtz, 2017.
NOTE: By *immediacy*, we mean direct and immediate involvement. By *compliance*, we refer to being in accordance with established guidelines or specifications yielding standardization. We refer to a group of organizations combined or making a joint effort to undertake some specific task as a *syndication*.

Findings and Recommendations

Air Force WRM is not truly globally managed today. Over the past decade, the Air Force has centralized some aspects of WRM support by establishing the 635th Supply Chain Operations Wing (SCOW). The SCOW maintains centralized acquisition, distribution, and inventory for a small number of Air Force assets, basic expeditionary airfield resources (BEAR), fuel support

---

[1] By *information distortions*, we mean information that is altered, omitted, or otherwise reorganized as it is transmitted between organizations.

[2] Paul R. Kleindorfer and Germaine H. Saad, "Managing Disruption Risks in Supply Chains," *Production and Operations Management*, Vol. 14, No. 1, Spring 2005.

[3] Thomas W. Malone, "Making the Decision to Decentralize," Harvard Business School Working Knowledge, March 29, 2004.

equipment, and other support equipment. Even for those assets, the SCOW does not control positioning or maintenance, and it has no oversight of WRM beyond those classes of supply. The current system is positioned more for efficiency than for effectiveness.

The CDO environments expected in the future will likely create new risks and uncertainty, and such environments might be better served through a centralized management approach. Air Force munitions management processes provide an example of a centralized or global management approach. The stakeholders work together to prioritize the acquisition, allocation, and positioning of munitions, globally. One of the advantages of this centralized management approach is that it is a transparent process that all the stakeholders understand. Because this process is well understood, we did not look at munitions management in our analysis. Establishing the 635 SCOW with WRM responsibility and authority is a step toward centralization; however, there are process, analytic capability, tool, and system enhancements that could further centralize the management of WRM (with decentralized execution) to better support the U.S. strategic goals of resiliency and agility.

In this analysis, we consider classes of WRM beyond those currently managed by the SCOW. We offer some recommendations for methods that a global manager of WRM could employ to improve global WRM management. First, **a global manager of WRM should standardize and validate warfighter demands.** The Air Force has some systems in place today to validate some but not all functional requirements (for example, BEAR; munitions; and petroleum, oil, and lubricants). Such tools as the Strategic Tool for the Analysis of Required Transportation (START) and Lean-START (see Figure S.2), developed by RAND Project AIR FORCE (PAF), could be used by a global manager to generate and compare requirements across areas of responsibility.[4]

---

[4] See Don Snyder and Patrick Mills, *Supporting Air and Space Expeditionary Forces: A Methodology for Determining Air Force Deployment Requirements*, Santa Monica, Calif.: RAND Corporation, MG-176-AF, 2004; and Patrick Mills, James A. Leftwich, Kristin Van Abel, and Jason Mastbaum, *Estimating Air Force Deployment Requirements for Lean Force Packages: A Methodology and Decision Support Tool Prototype*, Santa Monica, Calif.: RAND Corporation, RR-1855-AF, 2017.

Figure S.2. Inputs in the Lean-START Model, in Which the User Defines Characteristics

Duration	
Total Duration (days)	10
Max flying days	1
Maintenance capability	Launch & recover

Aircraft	Number	Mission	Sortie rate	ASD	Sorties/day	Personnel	Weight (Stons)
B-2			-		0		
B-52			-		0		
F-16CD	12	Air-to-Ground	1.50	1.00	18	111	66
F-15C	0	No Munitions	-	-	0		
F-15E	0	No Munitions	-	-	0		
F-15J	0	No Munitions	-	-	0		
FA-18	0	No Munitions	-	-	0		
F-22	0	No Munitions	-	-	0		
C-130	0		-		0		
EC-130	0		-		0		
MC-130	0		-		0		
HC-130	0		-		0		
AC-130	0		-		0		
KC-130	0		-		0		
KC-10	0		-		0		
KC-135	0		-		0		

Base Input Parameter Options

Base_Type	Conventional Threat	CBRN Threat
High Capability (MOB/COB)	High	High
Med. Capability (COB/IAP)	Low	Medium
Low Capability (COB)		Low
Austere		

POL Supply	Security_Forces	
	Protect base	EOD
Austere	Protect airfield	Threat-driven
Storage/Fuel provided	Protect flightline	Protect aircraft only
Fuel + Filstands provided	SF Team	Minimal
Full service provided	None	None

POL Support	Firefighting	CBRN Capability
	Protect aircraft and structures	Threat-driven
FORCE + filstands	Protect aircraft only	Minimal
FORCE + trucks	Extract pilot only	None
ABFDS	C2 only	
FARP	No support	Advanced ADR
		None
		Small
		Medium
		Large
		Very large

Lean-START determines the type and amount of equipment and personnel required to support a user-defined operation. The tool can be used as a demand generator and iterative planning environment for different courses of action and risk assessments. We use Lean-START to demonstrate the type and utility of a tool needed by a global manager of WRM to validate warfighter requirements beyond the few individual systems such managers currently have.

The Air Force should move away from inventory management and toward capability management for mission support assets. We created buckets of WRM capabilities aligned with a base's different activities over different periods of time. The categories, which are ordered based on how quickly each is needed (*temporal order*), are as follows:

- *Category 1: Open the base and prepare to receive forces*; example capabilities include material handling equipment, force protection, and communications
- *Category 2: Force projection*; example capabilities include maintenance, sortie generation, fuels, munitions, crash rescue, and explosive ordnance disposal
- *Category 3: Base operating support (BOS) initial operating capability (IOC)*; example capabilities include minimal housekeeping, a special operations–like capability, power production, and limited industrial BEAR
- *Category 4: Recover the base and aircraft*; example capabilities include rapid airfield damage recovery and aircraft battle damage repair
- *Category 5: BOS full operating capability (FOC)*; example capabilities include full housekeeping and general-purpose vehicles.

Capability buckets provide a way for a global manager to think about which individual functions should be grouped together to form operational capabilities and then how those capabilities need to be prepositioned.

Once a global manager knows which capabilities need to be prepositioned, he or she needs to prioritize them. Thus, **the Air Force should develop a method to determine priority for WRM prepositioning.** To assess prioritization trade-offs, we developed a decision tree that evaluates the temporal order of the WRM capability and aligns that order (as well as other aspects, such as transportability and cost) to a prepositioning posture (see Figure S.3). This decision tree is an example of the type of repeatable process that a global manager could use to think about how to posture the most time-critical WRM capabilities and which posture could be used for less-critical capabilities, such as those available for purchase in the local economy.

Figure S.3. Overview of the Decision Tree Framework to Provide Posture Options

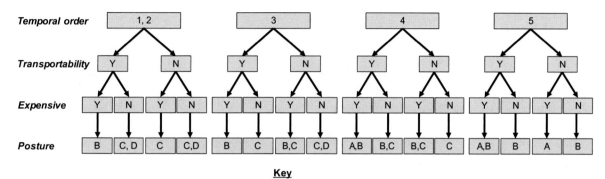

Key

Mission Criticality
1. Prepare to receive forces
2. Force projection
3. BOS IOC
4. Recover the base / aircraft
5. BOS FOC

Posture
A. Centralized in nearby theater or CONUS
B. Centralized in theater
C. Moderately distributed in theater
D. Highly distributed in theater

NOTE: CONUS = continental United States.

The Air Force should optimize its WRM prepositioning posture. PAF researchers developed a tool, the Prepositioning Requirements Planning Optimization (PRePO) model,[5] to evaluate WRM posture options. The model was designed to find optimal solutions—for example, for lowering costs, minimizing the time it takes for a base to receive all of its required WRM, or evaluating user-developed postures—based on storage allocations among candidate sites, costs of storage and transportation, and a transportation schedule. In this analysis, we used the PRePO model to demonstrate the types of analyses that a global manager of WRM should be conducting. As an example, we assessed the cost and global performance of a set of WRM storage courses of action for conflicts in four different global regions.

As a notional example, Figure S.4 shows the effect of a moderately dispersed WRM storage posture versus a centralized storage posture. Both storage postures have the same amount of transportation assets available. By day 13, all forward operating locations (FOLs) in the more dispersed posture have received the WRM required for IOC—four days earlier than the centralized storage course of action with the same lift available. Decision support tools like the

---

[5] The PRePO model and its application to previous analyses are discussed in detail in Brent Thomas, Bradley DeBlois, Katherine C. Hastings, Beth Grill, Anthony DeCicco, Sarah A. Nowak, and John A. Hamm, *Developing a Global Posture for Air Force Expeditionary Medical Support*, Santa Monica, Calif.: RAND Corporation, forthcoming, Not available to the general public.

PRePO model can show decisionmakers the costs and risks associated with various positioning strategies.

Figure S.4. Region A Bases Receiving the WRM Required for IOC over Time

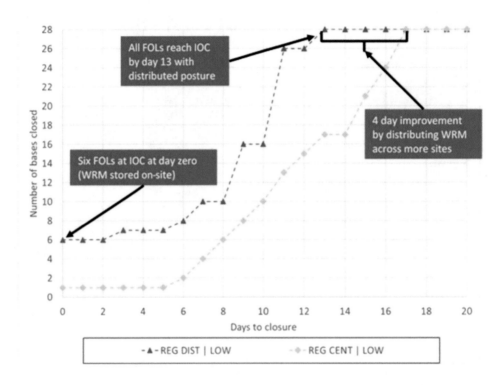

NOTE: REG DIST | LOW = WRM storage is moderately distributed across the region, and availability of transportation assets is low; REG CENT | LOW = WRM storage is centralized regionally, and availability of transportation assets is low.

To supplement the PRePO analysis, we developed the Strategic and Political Risk Assessment Tool, which considers political, economic, and strategic factors beyond what is modeled in the PRePO tool. Such qualitative factors as a host nation's long-term relationship with the United States and its internal political and economic stability can be informative when thinking about where to preposition WRM assets. **The Air Force should consider political factors when analyzing WRM prepositioning strategies**. A planner or analyst can use the political considerations to influence the inputs into the PRePO model.

Finally, we recommend that the Air Force **adopt tools and metrics to measure readiness for mission support assets**. As found by the Air Force's Sustainment Review Team in 2018,[6] reporting on WRM equipment readiness is inconsistent across commands and may not always accurately reflect equipment condition. The Air Force reports aircraft and supply support

---

[6] In 2018, the Air Force established a Sustainment Review Team to evaluate the readiness of Air Force parts, supply, and equipment. The team identified several issues, including one about reporting WRM equipment readiness (the report is not available to the general public).

readiness in the Status of Resources and Training System (SORTS). But the Air Force needs a way to regularly report its ability to provide mission support capabilities, not item inventories—perhaps in the SORTS or the Defense Readiness Reporting System.

To achieve true global management of WRM, with the capabilities recommended here, the Air Force needs to invest in an organization with adequate resourcing (manpower, facilities, tools, and systems), authorities, and governance to buy WRM capabilities, sustain them, and make trade-off decisions about their global positioning. This would be a cultural shift from how WRM is managed today.

Acknowledgments

Numerous people inside and outside of the U.S. Air Force provided valuable assistance to and support of this analysis. In particular, we thank Lt Gen John Cooper, former Deputy Chief of Staff for Logistics, Engineering, and Force Protection (AF/A4); Lt Gen Warren Berry, current AF/A4; Lt Gen Lee Levy, former commander of the Air Force Sustainment Center (AFSC/CC); and Lt Gen Donald Kirkland, current AFSC/CC. We also thank, from the AF/A4 staff, Col Luther King, Lt Col Andrew Marsiglia, and William Sweet, for their continued support throughout the analysis.

This work would not have been possible without support from the 635th Supply Chain Operations Wing (SCOW), especially Col David Sanford, former commander of 635 SCOW; Col Robert Henderson, current commander of 635 SCOW; Mark Hawley, vice director of 635 SCOW; and Rex Lutz, WRM Global Management Office. We also thank the staffs of Pacific Air Forces and of U.S. Air Forces in Europe and Air Forces Africa for their time. We appreciate their willingness to assist us throughout the project.

We also thank Brianna Shipp, U.S. Naval Supply Systems Command, Office of Corporate Communications, and Armin Zotaj, Director Navy Staff/Directive, Forms, and Report Management, for their help with understanding naval WRM.

Finally, we thank those at the RAND Corporation who helped us improve this work. We are especially grateful to Dan Norton and Don Snyder for their careful and thoughtful reviews of this report. We are also thankful for the support given by Caroline Baxter, Josh Girardini, Beth Grill, Jennifer Kavanaugh, Nirabh Koirala, Bradley Martin, Stacie Pettyjohn, Laura Poole, James Powers, Michael Ryan, Ryan Schwankhart, Anita Szafran, David Thaler, Brent Thomas, and Alan Vick.

As always, the analysis and conclusions are solely the responsibility of the authors.

Abbreviations

A4	Logistics, Engineering, and Force Protection
AF/A4	Air Force Deputy Chief of Staff for Logistics, Engineering, and Force Protection
AF/A4LR	Air Force Deputy Chief of Staff for Logistics, Engineering, and Force Protection, Logistics Readiness Directorate
AF/A4P	Air Force Deputy Chief of Staff for Logistics, Engineering, and Force Protection, Resource Integration Directorate
AF/A5R	Air Force Deputy Chief of Staff, Plans and Programs, Operational Capability Requirements Directorate
AF/A9F	Air Force Deputy Chief of Staff for Studies, Analysis, and Assessments, Force Structure Analyses Directorate
AFCEC	Air Force Civil Engineer Center
AFI	Air Force Instruction
AFMC	Air Force Materiel Command
AOR	area of responsibility
APF	afloat prepositioning fleet
APS	Army prepositioned stocks
BEAR	basic expeditionary airfield resources
BOS	base operating support
CCDR	combatant commander
CDO	contested, degraded, or operationally limited
CJCSI	Chairman of the Joint Chiefs of Staff Instruction
COA	course of action
CODE	combat operations in denied environments
COMAFFOR	commander of Air Force forces
CONOP	concept of operation
CONUS	continental United States
DABS	deployable air base system

DoD	U.S. Department of Defense
DRRS	Defense Readiness Reporting System
EVI	Environmental Vulnerability Index
FEMA	Federal Emergency Management Agency
FOC	full operating capability
FOL	forward operating location
FORCE	fuels operational readiness capability equipment
FSL	forward support location
FY	fiscal year
GCC	geographic combatant command
GDP	gross domestic product
GLOB DIST	storage is distributed globally
GMO	global management office
HQDA	Headquarters, Department of the Army
IOC	initial operating capability
LOGMOD	Logistics Module
LPI	Logistics Performance Index
MAJCOM	major command
MCO	major combat operation
MMG	Materiel Maintenance Group
NATO	North Atlantic Treaty Organization
OPNAV	Office of the Chief of Naval Operations
PACAF	Pacific Air Forces
PAF	Project AIR FORCE
POL	petroleum, oil, and lubricants
PRePO	Prepositioning Requirements Planning Optimization
PWRM	prepositioned WRM
REG CENT	storage is centralized regionally
REG DIST	storage is moderately distributed across the region

RSP	readiness spares package
ROBOT	RAND Overseas Basing Optimization Tool
SCOW	Supply Chain Operations Wing
SEATO	Southeast Asia Treaty Organization
SME	subject matter expert
SORTS	Status of Resources and Training System
SPRAT	Strategic and Political Risk Assessment Tool
START	Strategic Tool for the Analysis of Required Transportation
TPFDD	time-phased force deployment data
USAFE	U.S. Air Forces in Europe
UTC	unit type code
WRM	war reserve materiel

Chapter One

Introduction and Analytic Approach

The operational environment of the future may be different from the environment that the U.S. military has been accustomed to over the past 30 years. Adversaries have developed capabilities that may restrict or deny U.S. forces' access into and within areas of operation. Potential threats from Russia, China, Iran, and North Korea include both kinetic threats, such as cruise and ballistic missiles, and non-kinetic threats, such as cyberattacks, and could result in the U.S. military operating in a contested, degraded, or operationally limited (CDO) environment. Prepositioning select war reserve materiel (WRM) may help mitigate vulnerabilities associated with operating in a CDO environment, and the WRM may serve as a signal of U.S. presence in an area of responsibility (AOR) and thus help deter aggression and coercion by adversaries.

The objective of this analysis is to evaluate management approaches and global prepositioning strategies for WRM postures for AORs in CDO environments. Currently, the 635th Supply Chain Operations Wing (SCOW) is the Air Force WRM global manager; however, the SCOW manages only three Air Force commodities—basic expeditionary airfield resources (BEAR), fuel support equipment, and other support equipment. In this analysis, we consider a full spectrum of WRM. We do not, however, include munitions.[1]

In this report, we describe conditions under which global management practices are advantageous and then propose methods that a global manager of WRM could employ to improve support of air component operational warfighting demands. Specifically, we demonstrate ways to standardize and validate determination processes for WRM requirements, establish a WRM prioritization schema, relate WRM priority to positioning postures, analyze trade-offs through modeling, and assess partner-nation risk.

Approach

There are different ways to manage WRM in a resource-constrained environment, and each has associated risk. For example, the U.S. Air Force could contract U.S. or other contractors for performance-based logistics. These performance-based contracts would need to contain surge clauses and require delivery of WRM at time of need. Or the Air Force could continue to employ its current WRM structure, with decentralized computation of requirements, some decentralized processes (WRM maintenance and storage), and centralized execution for a few commodities. A

---

[1] Munitions WRM processes are different from the other classes of supply. The Global Ammunition Control Point, the global manager for munitions, has well-defined processes tied to operationally relevant metrics. However, even though the munitions processes could serve as an example of how to globally manage a commodity, they are not integrated with other functional areas or capabilities. See Chapter Three for more details.

third management strategy is global management, in which processes are controlled centrally, with decentralized execution for the core functions. For WRM, core functions include acquisition, storage and maintenance, distribution, and assessments. In this analysis, we focus on a global management strategy.

The analytic approach used in this analysis is shown in Figure 1.1. We began our effort by reviewing prior analyses of WRM and CDO environments. We then conducted an academic literature review of global management. We investigated the use of global management in industry and evaluated under which conditions global management of WRM could provide better support to the warfighter. We reviewed other armed services' WRM practices and then compared current Air Force WRM practices with how WRM could be globally managed. Finally, in this report, we suggest processes, analytic capabilities, tools, and systems that the Air Force could employ to improve global management of WRM.

Figure 1.1. Analytic Approach

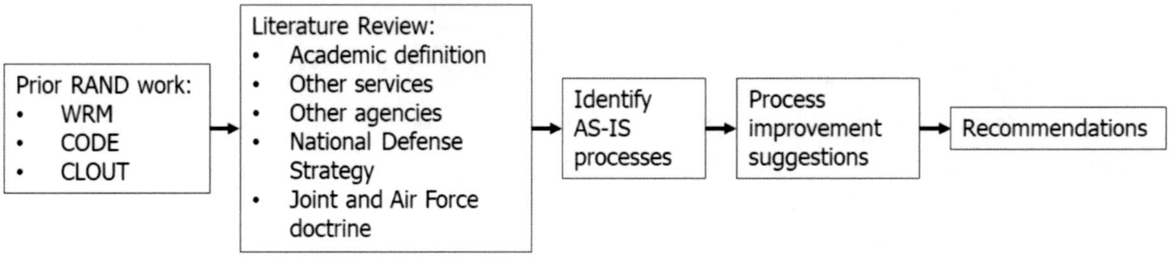

NOTE: CLOUT = coupling logistics with operations to meet uncertainty and the threat; CODE = combat operations in denied environments.

As we collected information and data for the analysis, we met with many organizations and consulted many sources. In particular, from within the Air Force, we consulted personnel from

- Air Force Deputy Chief of Staff for Logistics, Engineering, and Force Protection (AF/A4)
- Air Force Sustainment Center (AFSC)
- U.S. Air Forces in Europe and Air Forces Africa
- Pacific Air Forces (PACAF)
- Seventh Air Force (7 AF)
- Air Force Special Operations Command (AFSOC)
- 635 SCOW
- 635th Materiel Maintenance Group (635 MMG).

In addition, we consulted information and personnel from other organizations, such as

- U.S. Army
- U.S. Navy
- U.S. Marine Corps
- U.S. Coast Guard

- Commercial drilling industry
- Federal Emergency Management Agency (FEMA).

Organization of This Report

This report has six chapters. In Chapter Two, we present background information on previous WRM analyses and discuss global management from an academic perspective. Chapter Three outlines current Air Force WRM processes and management practices. In Chapter Four, we suggest methods that could be employed by a global manager of Air Force WRM. In Chapter Five, we present the Strategic and Political Risk Assessment Tool (SPRAT). The conclusions and recommendations are presented in Chapter Six.

The report also has four appendixes. Appendix A describes five case studies of global management outside the U.S. Air Force. Appendix B provides a network analysis of current Air Force WRM processes. Appendix C includes sample analyses using a prepositioning optimization model developed by RAND Project AIR FORCE (PAF). And Appendix D contains the SPRAT coding scheme and data sources.

Chapter Two
A Review of WRM and Global Management

In this chapter, we discuss prior WRM analyses and global management practices. We begin with a summary of the prior PAF analyses of WRM. Next, we discuss what academic literature says about global management usage and practices.

Prior RAND Analyses of WRM

PAF has a long history of analyzing WRM processes and practices. In a previous study, PAF researchers developed an analytical framework and optimization model (later called the RAND Overseas Basing Optimization Tool, or ROBOT) that minimized forward support location (FSL) operating, construction, and transportation costs needed for deterrence and training exercises while maintaining the storage capacity and throughput needed to win major regional conflicts and small-scale contingency operations.[1] The optimization model identified a set of WRM allocations across a series of potential FSL sites that would minimize the peacetime costs of deterrence and training missions while fulfilling time-phased operational requirements should deterrence fail.[2]

A subsequent study used this framework to evaluate the ability of the Air Force's existing overseas combat support basing and global WRM prepositioning postures to satisfy operational requirements across alternative future scenarios and deployment timelines. The authors evaluated the benefits and costs of a global posture of potential FSL sites against the Air Force's allocation of munitions, rolling stock, and basic expeditionary airfield resources (BEAR) at existing FSL sites.[3] The expanded set of existing and potential FSL sites identified by the optimization model offered the same capabilities as existing FSLs at a reduced overall cost to the Air Force, achieving as much as a 30-percent savings in total costs in the most likely baseline future scenario.[4]

As a result of these findings, the authors recommended that the Air Force adopt a global approach for selecting overseas combat support bases as a more effective and efficient

---

[1] Mahyar A. Amouzegar, Robert S. Tripp, Ronald G. McGarvey, Edward W. Chan, and C. Robert Roll, Jr., *Supporting Air and Space Expeditionary Forces: Analysis of Combat Support Basing Options*, Santa Monica, Calif.: RAND Corporation, MG-261-AF, 2004.

[2] Amouzegar et al., 2004, pp. 51–64.

[3] Mahyar A. Amouzegar, Ronald G. McGarvey, Robert S. Tripp, Louis Luangkesorn, Thomas Lang, and Charles Robert Roll, Jr., *Evaluation of Options for Overseas Combat Support Basing*, Santa Monica, Calif.: RAND Corporation, MG-421-AF, 2006.

[4] This savings was generated largely by a reduction in overall transportation costs to the system as a result of adding five potential FSLs to the existing 11 FSL sites. See Amouzegar et al., 2006, p. 69.

alternative to allocating resources on a regional basis. The authors also found that, although storing WRM and munitions aboard afloat prepositioning fleets (APFs) would offer additional flexibility and reduce vulnerabilities associated with land-based storage, APF storage was a much more expensive alternative to land-based storage and introduced its own risks to deployment timelines, particularly in light of the limited number of ports capable and willing to receive large APF cargo ships even with generous advance warning.[5] The study also assessed the political and operational benefits and drawbacks of a wider set of potential locations, and the authors recommended that the Air Force employ multimodal transportation options—including a combination of air, land, and sealift—to achieve a rapid logistics response with limited transportation assets.[6] At the time, the Air Force chose not to implement these recommendations, given the reluctance of regional combatant commanders (CCDRs) to transfer authority of their WRM assets to a global ownership and management system. Policymakers also expressed concerns over making the large investments in infrastructure necessary for a global posture.[7]

In fiscal year (FY) 2006, PAF was asked to revisit global WRM prepositioning and extend existing optimization models to incorporate some factors not addressed in previous work, including alternative packaging configurations for BEAR and WRM vehicles,[8] as well as different maintenance and transportation options for these assets, including the Virtual Afloat

---

[5] Amouzegar et al., 2006, pp. xxv, 45–46, 61–64. The costs of leasing, operating, and maintaining APF ships are significantly greater than the costs of land-based WRM storage. The reduction in transportation costs achieved by using APF ships was offset by a considerable increase in APF leasing and operating costs. When optimizing WRM allocation across the existing set of FSL locations, the ROBOT model did not select any APF munitions ships for storage, finding instead that it would be more cost-effective to upgrade and expand facilities at some of the 11 existing FSL sites identified as meeting the operational demands of the baseline scenario.

[6] Amouzegar et al., 2006, pp. 81–86. Several earlier RAND reports demonstrated that overreliance on airlift may reduce response capability because of throughput constraints and airlift availability (see, for example, Amouzegar et al., 2004; and Alan Vick, David Orletsky, Bruce Pirnie, and Seth Jones, *The Stryker Brigade Combat Team: Rethinking Strategic Responsiveness and Assessing Deployment Options*, Santa Monica, Calif.: RAND Corporation, MR-1606-AF, 2002). The ROBOT model was able to make better use of trucks and high-speed sealift in the expanded set of potential FSL sites, yielding a roughly 50-percent reduction in airlift usage without compromising operational requirements. Using the expanded set of global FSL options also reduced overall reliance on transport in light of the collocation of storage sites at forward operating locations (FOLs). See Amouzegar et al., 2006, p. 75.

[7] Ronald G. McGarvey, Robert S. Tripp, Rachel Rue, Thomas Lang, Jerry M. Sollinger, Whitney A. Conner, and Louis Luangkesorn, *Global Combat Support Basing: Robust Prepositioning Strategies for Air Force War Reserve Materiel*, Santa Monica, Calif.: RAND Corporation, MG-902-AF, 2010, pp. 4–5.

[8] This study considered two alternative packaging configurations for BEAR assets. Traditionally, BEAR assets are stored in *palletized* configurations that are designed for airlift but less well-suited for ground or sea transport. The modified ROBOT model analysis weighed this packaging against *containerized* configurations that store BEAR assets in shipping containers. Storing BEAR assets in containerized configurations reduces both vehicle loading and facility space requirements because these containers can easily be transferred to surface vessels and do not require inside warehouse storage (McGarvey et al., 2010, pp. 28–29). The authors also analyzed costs associated with different packaging alternatives for active and inactive WRM vehicles. Whereas active WRM vehicles are expected to be ready to support incoming forces, inactive WRM vehicles are stored in a *deep-storage* configuration at forward locations that would still allow them to be shifted quickly to mission-capable status. In 2003, PACAF began testing a deep-storage packaging configuration that wrapped inactive WRM vehicles in plastic (McGarvey et al., 2010, pp. 59–61), but this practice has since been discontinued. See footnote 13.

concept.[9] The resulting analysis employed a modified ROBOT optimization model to develop a global WRM prepositioning posture that satisfied deployment effectiveness constraints while minimizing total costs associated with time-phased deployment requirements for non–major combat operations (MCOs) and lesser contingencies and exercises.[10] Whereas previous prepositioning decisions had been optimized for predictable costs, the modified ROBOT model optimized against a wider set of predictable and contingency-dependent costs (what McGarvey et al., 2010, refers to collectively as *total system costs*) and allowed for potential reductions in transportation costs by placing FSL storage sites closer to FOLs for contingencies other than MCOs.[11]

In addition to employing a more widely geographically dispersed FSL posture, the total optimization model relied on alternative packaging, maintenance, and transportation options for BEAR and WRM vehicles. The modified ROBOT model also achieved reductions in maintenance and facility usage costs by shrink-wrapping WRM vehicles in plastic, but this practice has since been discontinued.[12] Furthermore, the total cost optimization model makes extensive use of traveling maintenance teams and, in the case of BEAR assets, asset swaps as cost-reducing alternatives to permanent on-site maintenance options that require greater investments in infrastructure and personnel.[13]

---

[9] The *Virtual Afloat concept* refers to the storage of a small amount of WRM assets in shipping containers that are prepositioned at or near shipping ports. These shipping containers can then be transferred to sea vessels and transported to FOLs to support deployment requirements. In addition to offering the benefits of containerized packaging configurations described earlier, the Virtual Afloat concept achieves many of the benefits of afloat storage on APF ships without incurring the large costs associated with leasing and operating APF vessels during periods in which their assets are not being transported in support of deployments. See McGarvey et al., 2010, pp. 5, 94.

[10] The results of this model can be best understood as a constant-effectiveness, variable-cost, and variable-risk analysis. See McGarvey et al., 2010, pp. xiii, 10.

[11] McGarvey et al., 2010, p. 11. Although the authors refer to the set of predictable costs and contingency-dependent costs optimized by the modified ROBOT model as *total costs* or *total system costs*, these terms do not incorporate other costs beyond these two categories. For example, because MCOs are funded through a separate supplemental funding process, the modified ROBOT model excludes the costs associated with supporting MCOs in its total cost analysis. See McGarvey et al., 2010, pp. 10–11.

[12] PACAF no longer shrink-wraps any vehicles or equipment because it found the practice not to be cost-effective. Sharp edges would cause the shrink-wrap to tear in harsh environments, allowing moisture to enter. (Even absent tearing, condensation would form on the interior of the wrapping, leading to corrosion.) Additional manpower was also required to unwrap vehicles that required maintenance or troubleshooting for routine problems, such as fluid leakage. The latest guidance issued by PACAF in 2012 regarding the maintenance and storage of WRM did not include any mention of previous instructions for shrink-wrapping WRM vehicles (Department of the Air Force, *Air Force War Reserve Materiel (WRM) Policies and Guidance*, Air Force Instruction 25-101, January 14, 2015). For PACAF's initial rationale for testing the shrink-wrapping of vehicles and its cost-saving potential, see McGarvey et al., 2010, pp. 59–61, 100.

[13] Traveling maintenance teams travel from FSLs with permanent on-site maintenance capabilities to FSLs that lack such a permanent capability. Similarly, asset swaps transport materiel requiring maintenance out of FSLs that lack permanent on-site maintenance capabilities and replace them with serviceable assets from FSLs that have such capabilities. See McGarvey et al., 2010, pp. 26, 100.

The authors further assessed the robustness and reliability of an optimized global FSL posture to continue to satisfy operational requirements in the event of disruption to FSL network facilities or other unexpected contingencies.[14] And analyses compared the results of employing an optimized global WRM management posture that allowed for cross-AOR support, wherein assets from one AOR could be moved without additional delay to an outside FOL, against an optimized alternative that did not allow for any cross-AOR transfer. The global WRM management posture allowing cross-AOR support achieved large cost reductions, mostly because a global pool of WRM resources reduces procurement, storage, operating, and maintenance costs associated with AOR ownership of WRM assets.[15]

More recently, PAF researchers have continued to evaluate the effectiveness and efficiency of WRM prepositioning postures to support combat operations in CDO environments. Since 2012, PAF has conducted a body of research evaluating combat operations in denied environments (CODE). CODE research has focused on BEAR, fuel support equipment, other support equipment, rapid airfield damage recovery, shelters, and munitions. Over seven years, CODE analyses quantified potential damage to air bases across a range of attack vectors and identified large requirements for shelters and rapid airfield damage recovery assets and materials.[16] The analyses further developed regional, cost-optimized WRM prepositioning strategies for both U.S. European Command and U.S. Indo-Pacific Command AORs.

In 2017, PAF researchers developed a decision support tool prototype designed to assist combatant command planning staffs in identifying combat support packages that can satisfy operational demands while reducing the speed of deployments. Extending PAF's earlier Strategic Tool for the Analysis of Required Transportation (START),[17] the Lean-START prototype allows planners to evaluate the trade-offs between speed and risk tolerance as they estimate the size of the combat support footprint (a sum of the amount of personnel and equipment) they will need

---

[14] *Robustness* in this context refers to the ability of a global WRM prepositioning posture to meet deployment demands across a variety of uncertain and potential contingencies. *Reliability* refers to the capacity of a global WRM prepositioning posture to satisfy deployment requirements in the event of disruption to network facilities, such as a denial of access to one or more FSL sites.

[15] McGarvey et al., 2010, pp. 100–101. Whereas the reliable optimization model mentioned earlier optimized global WRM prepositioning across short-term disruption to single FSL sites, an additional RAND study developed a series of models identifying the costs necessary to sustain WRM prepositioning postures that can endure multiple disruptions for more-extended durations. See Thomas Lang, *Defining and Evaluating Reliable Options for Overseas Combat Support Basing*, Santa Monica, Calif.: RAND Corporation, RGSD-250, 2009.

[16] Brent Thomas, Mahyar A. Amouzegar, Rachel Costello, Robert A. Guffey, Andrew Karode, Christopher Lynch, Kristin F. Lynch, Ken Munson, Chad J. R. Ohlandt, Daniel M. Romano, Ricardo Sanchez, Robert S. Tripp, and Joseph V. Vesely, *Project AIR FORCE Modeling Capabilities for Support of Combat Operations in Denied Environments*, Santa Monica, Calif.: RAND Corporation, RR-427-AF, 2015.

[17] Don Snyder and Patrick Mills, *Supporting Air and Space Expeditionary Forces: A Methodology for Determining Air Force Deployment Requirements*, Santa Monica, Calif.: RAND Corporation, MG-176-AF, 2004.

for deployments in CDO environments.[18] Whereas earlier planning concepts made static assumptions about needed levels of capability and duration for contingencies in CDO environments, the Lean-START prototype allows planners to adjust these inputs to generate lists of combat support packages tailored to the operational demands of specific CDO environments and the trade-offs the planners manage with regard to safety, security, and human performance and endurance.

In a separate study, PAF researchers explored five alternative courses of action (COAs) for the storage of 24 deployable air base system (DABS) "basic" sets to inform the prepositioning strategy of U.S. Air Forces in Europe (USAFE). Although a more widely dispersed prepositioning strategy would reduce transportation delays, the uncertainties surrounding availability and favorability of chosen airfields in a given contingency, coupled with the negligible differences in costs across the five alternative COAs, led the researchers to conclude that USAFE's current regional plan achieves the shortest transportation timelines at minimal risk of disruption.[19] This FY 2018 analysis built on previous assessments by incorporating a broader array of nonmaterial and qualitative factors that have not been systematically addressed in previous studies. These factors include methods to prioritize WRM, positioning strategies based on prioritization, political concerns, the inherent deterrence value of prepositioned WRM assets, and potential benefits and risks associated with greater cooperation with partner nations in fulfilling WRM storage, delivery, and combat support requirements.

Global Management in Academic Literature

The rapid expansion of global capital markets that facilitated more-fluid financial flows across borders, coupled with astonishing growth rates in developing countries, increasingly integrative technologies, and marked reductions in trade restrictions between nations, gave rise in the post–World War II era to an increasingly globalized economy in which international firms

---

[18] Using Lean-START, planners supply mission-specific inputs, such as the mission duration and the number and types of aircraft needed in a given operational scenario; specify a desired level of nonmaintenance agile combat support capabilities for beddown planning; and then tailor their expected support requirements according to the readiness of prepositioned assets, the availability of support from partner nations, and their risk tolerance level. The prototype generates an estimated footprint size according to a series of unit type codes (UTCs) tailored to the planner's input parameters. The planner can then adjust these parameters to generate different support options according to desired speeds of deployment and levels of risk accepted. See Patrick Mills, James A. Leftwich, Kristin Van Abel, and Jason Mastbaum, *Estimating Air Force Deployment Requirements for Lean Force Packages: A Methodology and Decision Support Tool Prototype*, Santa Monica, Calif.: RAND Corporation, RR-1855-AF, 2017.

[19] A storage strategy that prepositioned the DABS assets at six storage sites in Eastern Europe was more than two-thirds less expensive than the cost of storing these assets in CONUS when cost-sharing with partner nations was assumed to be unavailable. (This reduction in costs was driven largely by lower wage rates compared with those in other COAs.) However, the researchers assumed that USAFE would not concentrate all its support assets so close to potential threats, and removing this posture from consideration meant that the cost differences across all other European COAs were negligible. This study is detailed in unpublished 2018 RAND research, led by Patrick Mills, on the costs and benefits of potential USAFE prepositioning strategies.

confronted new challenges—and new opportunities—in competing in foreign markets.[20] By the 1980s, industry leaders and academic scholars at leading business schools began weighing the benefits and drawbacks of different management strategies that multinational enterprises could adopt to take advantage of new markets while minimizing the uncertainty and risks associated with competing on the international stage.

Understanding Global and Regional Management Strategies

A central tenet of the debate has focused on whether global management strategies or their more localized and regional counterparts can better equip multinational firms to compete effectively in meeting consumer demand while efficiently reducing overall costs. Firms that compete internationally must decide how to manage and geographically distribute the various activities involved in supplying goods and services to consumer markets, such as procurement, manufacturing, marketing, delivery, and research and development.[21]

In pursuing global or regional strategies, international firms vary across two dimensions: (1) the degree to which these value-added activities are geographically *concentrated* and (2) how tightly linked, or *coordinated*, the geographically dispersed entities that perform different functions are within a multinational enterprise.[22] The management strategy that a firm selects has significant implications not only for the concentration and coordination of subsidiary assets but also for the firm's overall organizational structure, which varies according to the degree of *centralization* of decisionmaking and management authority. Whereas centralized strategies concentrate decisionmaking authority in the hands of global managers, strategies that are more regionally oriented decentralize authority, assigning decisionmaking rights to more-autonomous

---

[20] Michael E. Porter, "Introduction and Summary," in Michael E. Porter, ed., *Competition in Global Industries*, Boston, Mass.: Harvard Business School Press, 1986b, pp. 1–5. There has been no shortage of work on globalization, its evolution in the 20th century, and its performance in the first decade of the 21st century. For a more academic treatment, see Jeffrey A. Freiden, *Global Capitalism: Its Fall and Rise in the Twentieth Century*, New York: W.W. Norton & Company, 2006. For more-popular treatments aimed at more-general audiences, see Thomas Freidman, *The World Is Flat: A Brief History of the Twenty-First Century*, New York: Farrar, Straus, and Giroux, 2005; and Fareed Zakaria, *The Post-America World*, New York: W.W. Norton & Company, 2008. For more-critical perspectives, see Joseph Stieglitz, *Globalization and Its Discontents*, New York: W.W. Norton & Company, 2002; and Dani Rodrick, *The Globalization Paradox: Democracy and the Future of the World Economy*, New York: W.W. Norton & Company, 2011.

[21] Michael E. Porter, "Competition in Global Industries: A Conceptual Framework," in Michael E. Porter, ed., *Competition in Global Industries*, Boston, Mass.: Harvard Business School Press, 1986a, pp. 19–23. Porter distinguishes between what he calls "primary activities," which consist of discrete tasks involved in the creation and delivery of a product or service (examples include inbound logistics, operations, outbound logistics, marketing, and sales) and "support activities," which provide inputs to facilitate the firm's primary activities (examples include human resource management; technological development; and procurement of raw materials, machinery, and infrastructure).

[22] Porter, 1986a, pp. 23–25.

regional subsidiaries that more directly deal with customers, competitors, frontline employees, and other stakeholders.[23]

A more localized or regional management strategy envisions a widely dispersed, loosely coordinated system of foreign subsidiaries empowered to develop differentiated products that are more closely tailored to local preferences. Firms pursuing such a strategy concentrate a substantial number of their activities in each country where they compete to respond more rapidly to shifts in consumer demand and reconfigure, distribute, market, and sell differentiated products accordingly. (In extreme cases, some firms will reproduce all segments of a supply chain close to market in each country where they operate.[24]) To facilitate this, firms adopt a decentralized organizational structure that delegates considerable operating independence and strategic decisionmaking authority to local or regional managers of foreign subsidiaries.[25] Central managers in multinational headquarters will coordinate financial flows and centralize some aspects of branding and product research and development, but decisionmaking authority is concentrated with regional managers, who are ultimately held responsible for results.[26] This enables these multinational firms to maintain a "national responsiveness" posture through which they tailor their products to suit the needs of particular markets.[27] Industries in which firms must cope with a high degree of differentiation among foreign markets include retail, insurance, and consumer finance.

A global management strategy, by contrast, envisions a tightly coordinated network of specialized foreign subsidiaries that can be geographically concentrated or widely dispersed, depending on the source of the firm's competitive advantage.[28] In contrast with those that supply products tailored to local demands of particular markets, international firms pursuing a global strategy specialize in the development of reliable, quality-controlled, standardized products that satisfy uniform preferences across foreign markets. To realize economies of scale, a foreign subsidiary specializes in the performance of only one aspect of supply chain production, exchanging its products or components with others in the network.[29] To implement this global strategy, firms adopt a centralized organizational structure, with largely one-way flows of goods, information, and resources from headquarters to a network of tightly coordinated and

---

[23] Herman Vantrappen and Frederic Wirtz, "When to Decentralize Decision Making, and When Not To," *Harvard Business Review*, December 26, 2017.

[24] Porter, 1986a, p. 25.

[25] Jahangir Karimi and Benn R. Konsynski, "Globalization and Information Management Strategies," *Journal of Management Information Systems*, Vol. 7, No. 4, Spring 1991, p. 11.

[26] Thomas Hout, Michael E. Porter, and Eileen Rudden, "How Global Companies Win Out," *Harvard Business Review*, September 1982, p. 103.

[27] Christopher A. Bartlett and Sumantra Ghosbal, "Managing Across Borders: New Strategic Requirements," *Sloan Management Review*, Vol. 28, No. 4, Summer 1987.

[28] Porter, 1986a, p. 29.

[29] Hout, Porter, and Rudden, 1982, p. 103.

interdependent subsidiaries. Although regional and local subsidiary managers exercise day-to-day control over their operations, key strategic decisions for worldwide operations are made centrally by senior management.[30] Firms that have moved toward global strategies include those in the commercial aircraft, semiconductor, electronics, and automotive industries.

In practice, firms rarely pursue strategies that are purely global or exclusively regional; more often, the strategies they pursue exhibit features of both. Business-oriented luxury hotel firms, such as Hilton, Sheraton, and Marriott, operate many geographically dispersed properties that are managed locally but retain the same brand names, formats, service standards, and worldwide reservation system.[31] In other cases, changes in the international market can compel firms that are organized around global strategies to adopt features of more regionally oriented strategies, and vice versa. In the automotive industry, for example, Toyota capitalized early on a relatively simple global strategy that concentrated production in a tightly linked, centrally coordinated network of interdependent subsidiaries and then distributed its standardized product worldwide. General Motors, by contrast, focused on a region-specific strategy that maintained geographically dispersed manufacturing facilities and separate brand names in different foreign markets. Increased tariff barriers incentivized Toyota to disperse its manufacturing operations more widely directly into foreign markets, and competitive pressures from Toyota's early success compelled General Motors to move toward a more global strategy.[32]

Benefits and Drawbacks of Global Management Strategies

Global management strategies offer multinational enterprises the ability to take advantage of economies of scale to produce standardized, quality-controlled products at reduced cost. A global strategy allows firms to exploit lower factor costs by locating segments of the supply chain, such as the manufacturing of different components, in low-cost countries with the comparative advantage in the performance of such activities, as has been the case with U.S. companies pursuing offshore manufacturing.[33] Global management strategies can also reduce inefficient redundancies in the supply chain that emerge among more-autonomous regional subsidiaries that maintain their own duplicated manufacturing, distribution, marketing, and servicing capabilities to deliver differentiated products and services country by country.

An interdependent network of manufacturers and suppliers also provides considerable operational flexibility to reallocate activities or resources in response to major disruptions to the supply chain or large shifts in exchange rates, tax rates, and transportation and labor costs.[34] This

---

[30] Karimi and Konsynski, 1991, p. 12.

[31] Karimi and Konsynski, 1991, p. 12.

[32] Porter, 1986a, pp. 28–29.

[33] George S. Yip, "Global Strategy . . . In a World of Nations?" *Sloan Management Review*, Vol. 31, No. 1, Fall 1989, p. 33.

[34] Yip, 1989, p. 33.

flexibility affords such firms greater leverage in adapting to the moves of global competitors. When a more centralized global firm faces a large competitor in a key market, it can funnel financial assets and other resources accumulated in one part of the world into the local market share battle through a strategy known as *cross-subsidization*.[35] Additionally, the coordinated set of centralized rules and procedures that global managers maintain across operating units ensures that the units adhere to uniform quality control and assurance standards.[36] Finally, the centralization of decisionmaking authority empowers global managers to resolve disputes among operating units and undertake hard decisions and tasks that subsidiaries may be unable or unwilling to do, such as invest in products or initiatives with distant or uncertain benefits or withdraw a product from a market entirely where its production may be profitable locally but inefficient globally.[37]

These global strategies, however, are not without their drawbacks. Standardized products at the endpoint of a globally managed supply chain may not be sufficiently tailored to local consumer demand to make them usable or desirable, much less profitable.[38] Moreover, the concentration of decisionmaking authority in the hands of centralized global managers at company headquarters can make it difficult for firms to detect and rapidly respond to fluctuations in consumer demand, the entry of a competitor to the market, or a supply chain disruption.[39] Although firms that maintain global strategies may be better equipped to respond to such disruptions more effectively in the medium to long terms because of their greater flexibility to reallocate resources across the supply chain, they will be slower to detect and respond to these changes in a more rapid time frame. Finally, restricting the autonomy and decisionmaking authority of local agents and subsidiaries can stifle motivation and inhibit the creativity that promotes innovation and problem-solving among frontline professionals, which can be especially limiting in high technology and service industries.[40]

When Should Organizations Adopt Global Management Strategies and Centralized Structures?

How do multinational corporations weigh these trade-offs in deciding to implement global or more regional, nationally responsive strategies? A firm's decision to centralize or decentralize

---

[35] Gary Hamel and C. K. Prahalad, "Do You Really Have a Global Strategy?" *Harvard Business Review*, July 1985, p. 144.

[36] George S. Yip, *Total Global Strategy II: Updated for the Internet and Service Era*, Upper Saddle River, N.J.: Prentice Hall, 2003, pp. 18–19.

[37] Vantrappen and Wirtz, 2017.

[38] Yip, 2003, p. 21.

[39] Yip, 1989, p. 34.

[40] Thomas W. Malone, "Making the Decision to Decentralize," Harvard Business School Working Knowledge, March 29, 2004.

management authority varies according to the nature of the product or service it offers in the industry in which it competes, as well as its assessment of the risks and uncertainties it seeks to mitigate. For firms that specialize in the delivery of locally tailored and highly differentiated products and services, a regional strategy that devolves greater autonomy to subsidiaries would likely be more profitable as suppliers adapt their products to be more nationally responsive to consumer demands, local laws, and economic regulations. By contrast, a global, more centralized management strategy would be more optimal for firms that develop and distribute a narrower set of reliable, standardized products and services whose quality is assured and whose servicing and maintenance can be routinized at relatively lower cost.[41]

The source and magnitude of the uncertainty and risks that a firm assesses it will confront will, in many ways, shape the management strategy it adopts to hedge against them. According to logistics scholars Ila Manuj and John T. Mentzer, a firm's exposure to risk can come from a variety of sources, including the following:

- *Supply risks:* disruption of supply, inventory, schedules, or technology access; price escalation; technological uncertainty; product changes; and frequency of design changes
- *Operational risks:* breakdown of operations and inadequate manufacturing or processing capabilities
- *Demand risks:* new product introductions by competitors and changes in demand and consumer preferences
- *Security risks:* security of infrastructure and information systems; vulnerability to terrorism, vandalism, crime, and sabotage
- *Macroeconomic risks:* economic shifts in wage rates, interest rates, exchange rates, and prices
- *Policy risks:* Actions taken by host governments, such as quota restrictions, the imposition of tariff barriers, and economic sanctions.[42]

A firm's assessment of its exposure to these sources of uncertainty can shape the kind of management strategy—global or regional—and the type of organizational structure—centralized or decentralized—that they adopt to mitigate against these risks. Firms that compete in industries susceptible to higher degrees of demand-side uncertainty are more likely to adopt more-regional management strategies that decentralize decisionmaking so that managers of regional subsidiaries can respond more quickly to fluctuations in consumer demand, which have the potential to change faster than the supply chain can adjust. Rapid shifts in localized demand give rise to information distortions whose adverse effects can reverberate back up the supply chain. In extreme cases, this can generate tremendous inefficiencies, a phenomenon known as the *bullwhip*

---

[41] Yip, 2003, p. 13.

[42] This list has been adapted slightly from Ila Manuj and John T. Mentzer, "Global Supply Chain Risk Management," *Journal of Business Logistics*, Vol. 29, No. 1, 2008, p. 138, Table 1.

effect.[43] Proponents of decentralized management strategies argue in favor of empowering regional or local subsidiaries to adjust to changes in demand with decisionmaking authority over production and distribution in ways that are tailored to the local market and thereby absorb distortions that would otherwise reverberate through the supply chain.[44]

Multinational firms that are more vulnerable to supply-side risks, such as disruptions to their supply chain, denial of access, or a breakdown of operations, may be better able to mitigate the effects of these adverse events by adopting a more centralized global management strategy. Firms that centralize decisionmaking and maintain a coordinated network of subsidiaries afford global managers the operational flexibility to adapt to major disruptions, such as political restrictions, natural disasters, or terrorist attacks, at one node by reallocating resources to other nodes within the supply network.[45] The centralization of too much decisionmaking, however, or a supply chain that is too tightly coupled can create its own vulnerabilities. Although efficiency may compel a globally managed firm to concentrate all of its manufacturing activities in a single location, the disruption of tightly coupled facilities to catastrophic events, such as natural disasters, major power outages, sabotage, or terrorist attacks, can grind operations to a standstill.[46] Too much complexity can introduce brittleness and vulnerabilities; therefore, firms would be well advised to incorporate agility, redundancy, and flexibility requirements into their global management strategies.

A firm pursuing a global management strategy may also be better equipped to handle the emergence of a strategic competitor that can challenge its market position globally. As discussed previously, although decentralized firms may be better informed to respond more rapidly to the entry of regional competitors into local markets, global managers may be better positioned to make difficult strategic decisions to close manufacturing plants or withdraw failing products from markets that are essential to the company's long-term viability against these competitors. Although it slows the firms' responsiveness to urgent national or regional-level developments, the detachment of global managers from the day-to-day or month-to-month shifts that consume the attention of mid-level managers grants them a vantage point to detect problems that fall outside the purview of any single operational unit and affords them a certain degree of freedom

---

[43] Hau L. Lee, V. Padmanabhan, and Seugjin Whang, "Information Distortion in Supply Chains: The Bullwhip Effect," *Management Science*, Vol. 42, No. 4, 1997.

[44] For similar arguments, see Alain Verbeke and Christian Giesler Asmussen, "Global Local, or Regional? The Locus of MNE Strategies," *Journal of Management Studies*, Vol. 53, No. 6, September 2016; Malone, 2004; and Alan M. Rugman and Alain Verbeke, "A Perspective on Regional and Global Strategies of Multinational Enterprises," *Journal of International Business Studies*, Vol. 35, No. 1, January 2004.

[45] Paul R. Kleindorfer and Germaine H. Saad, "Managing Disruption Risks in Supply Chains," *Production and Operations Management*, Vol. 14, No. 1, Spring 2005.

[46] Dia Bandaly, Ahmet Satir, Yasemin Kahyaoglu, and Latha Shanker, "Supply Chain Risk Management – I: Conceptualization, Framework and Planning Process," *Risk Management*, Vol. 14. No. 4, November 2012, pp. 254–257.

to overcome organizational barriers that inhibit sound strategic decisionmaking.[47] For example, a subsidiary operational unit that has a certain amount of excess inventory may be unwilling to commit its resources to assist its counterparts within the same organization out of concern for its own performance metrics.[48] A global manager with more-centralized decisionmaking is better equipped to overcome these barriers that would otherwise inhibit the reallocation of these resources.

Management Practices Summary

As part of our analysis, we conducted case studies of management practices in industry, a government organization (FEMA), and other armed services (see Appendix A), and we reviewed numerous publications about global management. From those reviews, we gleaned several overarching themes related to management practices (see Table 2.1). Most organizations choose a management strategy based on the organization's goals, anticipated risk, and uncertainty. Generally, organizations have a mix of management strategies—some centralized and some decentralized—tailored to support the activity that the organization serves. A decentralized management strategy facilitates a rapid response to changes in demand but can generate informational distortions—in which data are changed, omitted, or reorganized as they are communicated to other parts of the organization. A decentralized management strategy can lead to cost inefficiencies and promote organizational fragmentation. A more centralized management strategy promotes information-sharing, process standardization, and resilience of the supply chain to potential disruptions at the expense of time and a tailored solution.

Table 2.1. Common Themes in Global Management Literature

When it is important to have . . .	It is usually advantageous to veer toward . . .	Because that solution enables and stimulates . . .
Responsiveness	Decentralization	Immediacy
Reliability	Centralization	Compliance
Efficiency	Centralization	Syndication

SOURCE: Adapted from Vantrappen and Wirtz, 2017.
NOTE: By *immediacy*, we mean direct and immediate involvement. By *compliance*, we refer to being in accordance with established guidelines or specifications yielding standardization. We refer to a group of organizations combined or making a joint effort to undertake some specific task as a *syndication*.

---

[47] Vantrappen and Wirtz, 2017.

[48] Hau L. Lee and Corey Billington, "Managing Supply Chain Inventory: Pitfalls and Opportunities," *Sloan Management Review*, Vol. 33, No. 3, Spring 1992, p. 70. See also Nicolai J. Foss and Peter G. Klein, "Why Managers Still Matter," *Sloan Management Review*, Vol. 56, No. 1, Fall 2014.

Although industry's top priority is usually profit, the Air Force has a different motivation: defending the United States and its interests. Because that is the top priority, with no room to fail, the Air Force is often required to be responsive, reliable, efficient, and resilient, making it difficult to discern which management practice is most advantageous. For the Air Force, there is a constant need to balance responsiveness and efficiency. However, considering the likely CDO environments of the future, there may be a need to build more resilience against potential disruptions into the WRM system through centralization. We evaluate the Air Force's current WRM structure and its challenges in the next chapter.

Chapter Three
Current Air Force Management of WRM

In this chapter, we discuss how the Air Force is currently managing WRM—a process that has seen many changes over the past decade. We begin with a network analysis of WRM processes and organizations. From there, we document our understanding of current processes, identify some existing shortfalls, and highlight some additional challenges associated with a CDO environment.

A Network Analysis of WRM Based on Air Force Instructions

The Air Force's current policies and practices for managing WRM involve numerous organizations and tasks that are complexly intertwined. We found that a simple review of policy and discussions with subject matter experts (SMEs) was insufficient to disentangle the tasks and organizational relationships. To more clearly reveal these tasks and relationships, we created a formal model of existing WRM management, as stated in policy documents, in the form of a network. The network model provides insights into who the actors are, what they do, how central they are, and with whom they interact. It does this in a visual format in which information is traceable back to policy (see the example visualization of WRM organizations and tools in Figure 3.1). The following discussion of current management practices uses this network model for insights. The model (a database and visualization tool) is described in detail in Appendix B.

Figure 3.1. Example Output of WRM Network Analysis Linking Organizations to Organizations by Systems and Tools

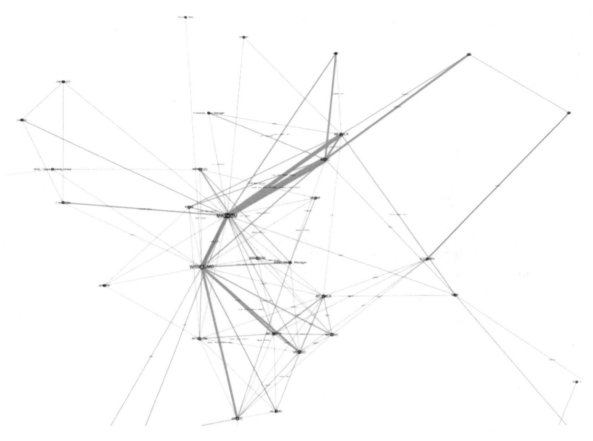

NOTE: Green represents the major command (MAJCOM) community; purple represents the WRM global manager community.

Current Management Practices

In this section, we describe our understanding of current WRM processes. This knowledge was derived from our review of Air Force Instructions (AFIs) and our network analysis, which together highlight the complexity of the WRM system and interrelatedness of its processes. Discussions with WRM SMEs further clarified how those processes, illustrated in the network analysis, are used in practice. We do not document the entire WRM system, because it is large and complex. Instead, we highlight the processes and functions for which we have suggested improvements in Chapter Five.

Supply Side

Air Force WRM is maintained, sustained, and stored by region. Therefore, the processes associated with WRM maintenance and sustainment vary across the AORs. Even for the pieces of WRM that are managed by the 635 SCOW (as the WRM global manager), sustainment and storage are the responsibility of the MAJCOM, not the SCOW. The current WRM storage and maintenance practices are centrally managed within the regions, gaining some regional efficiencies.

WRM is also managed by item or function, not by capability. For example, the 635 SCOW manages BEAR assets, but opening a base requires many more assets than just BEAR. Thus, multiple WRM managers are needed to produce a capability. From a mission support perspective, supporting an operating location requires BEAR, fuel support equipment, vehicles, communications, security forces, air traffic control, civil engineers, rations, mortuary affairs, medical, and other functions. These functions are managed by many different Air Force and other outside organizations using a variety of management approaches, some focused more on efficiency and others focused more on responsiveness. Table 3.1 lists some of those organizations and their functional responsibilities.

Table 3.1. Mission Support Capabilities, as Provided by Numerous Stovepiped Combat Support Functional Organizations

Organization	Function
Air Force	
Air Force Sustainment Center	Spares, BEAR, fuel support equipment, support equipment (Class VII)
Air Force Life Cycle Management Center	Munitions (Global Ammunition Control Point)
Air Force Installation and Mission Support Center	Rations, mortuary affairs, civil engineer, security forces
Air Combat Command, Communications Directorate	Communications
Air Force Medical Operations Agency	Medical
Other	
Defense Logistics Agency	Consumables, fuel, medical
General Services Administration	Some vehicles
Other services	Joint basing
Host or partner nations	Contracted support, some consumables, spares pool for F-35 aircraft

Air Force munitions management processes provide an example of a centralized or global management approach. The initial munitions requirements are generated by in-theater planners who identify the target sets needed to be prosecuted to achieve the CCDR objectives and the associated munitions to service those target sets. Air Force Deputy Chief of Staff, Plans and Programs, Operational Capability Requirements Directorate (AF/A5R) validates the requirement using the Combat Force Ammunition Model, which takes target sets and assigns primary, secondary, and tertiary munitions. From these analyses, AF/A5R can develop a future acquisition requirement. AF/A5R, the theater planners, and the Global Ammunition Control Point then work together to prioritize the acquisition, allocation, and positioning of the munitions globally. One of the advantages of this centralized management approach is that it is a transparent process that allows key stakeholders to understand the operational effects and allows any shortfalls and issues to be well understood by customers. And because this process is well understood by stakeholders, we did not look at munitions management in our analysis.

Even for functions that are globally managed, such as munitions, the many global providers are not integrated to provide information about mission support capabilities. Individual functional stovepipes can provide information on fill rates and functional capabilities; however, no one has an integrated view of how a constraint on a mission support capability could affect operational capabilities.

Demand Side

Global providers have limited visibility of operational requirements. Commander of Air Force forces (COMAFFOR) Logistics, Engineering, and Force Protection (A4) staffs work with COMAFFOR operational planners to establish requirements to meet CCDR operational intent in each AOR. WRM requirements are determined by region, with each staff developing requirements differently. The 635 SCOW holds a theater working group with each staff to review and validate WRM requirements for the functions for which the SCOW has responsibility. Other organizations, such as the Air Force Installation and Mission Support Center, attend the theater working groups but manage their WRM requirements separately. WRM requirements are compartmentalized by function instead of being integrated into the capability those functions provide to the warfighter.

On the other hand, A4 staffs have limited visibility into available global resources. They typically plan assuming that resources will be available when needed. However, without an integrated view of available mission support capabilities, staffs would have to check the availability of each item or function to determine whether resource shortfalls exist. COMAFFOR staffs lack the time, tools, and analytic capabilities to conduct large-scale assessments of disparate mission support functions.

Neither COMAFFOR A4 staffs nor global managers have the tools, manpower, or training to identify WRM resources that could be shared across theaters. They also cannot determine the effect that sharing those resources could have on operational requirements.

CDO Challenges That Affect WRM Management

As mentioned in Chapter One, adversaries can contest U.S. military operations, both kinetically and non-kinetically. A CDO environment is very different from the operational environment U.S. forces have faced over the past 30 years. The Air Force cannot assume that it will always have air superiority. A CDO environment creates additional challenges both for mission generation and for mission support.

A CDO environment is dominated by uncertainty and risk, particularly about the type and amount of damage the enemy may cause to U.S. military personnel and equipment. According to the academic literature, a CDO environment could be served by either a centralized management approach—to protect against disturbances of the supply chain—or a decentralized management approach—to improve responsiveness. When COMAFFOR staffs determine WRM requirements, they do not always plan and program for attrition from an attack. But even if they did account for WRM attrition, the specific damage and associated requirements will not be known until after the attack.

In a CDO environment, there is also risk that there may be additional requirements generated from dynamically changing concepts of operation (CONOPs), such as adaptive basing. As the U.S. military presses an operational advantage or retreats from an adversary attack, there could be additional WRM requirements. Depending on the CONOP, whether it was planned or not, those additional requirements may not be established until the CONOP is employed. Financial limitations might make investing in WRM capabilities to support a wide range of potential CONOPs across every AOR infeasible.

Functional stovepipes further complicate WRM investments. The Air Force currently does not have the ability to identify the most-binding constraint or to balance between or across functional stovepipes, making it difficult to determine where to invest the next WRM dollar. Even if a global manager could look across functional stovepipes to invest in the most-binding constraint, projected funding is not always executed as planned. The Air Force does not have a way to discern how a reduction in WRM funding would translate to operational capability—for example, the ability to open a base.

The 2018 National Defense Strategy

The 2018 National Defense Strategy set a new tone for the U.S. military.[1] The document clearly identifies regional priorities and states that all investments will be focused on those theaters. All other requirements for other theaters will be lesser included requirements. This is a shift from prior guidance. The National Defense Strategy further states that the U.S. military should focus on light, lean, and lethal capabilities; be prepared to move quickly and often; and prepare for short engagements. This is a distinct shift from the extended engagements in which the U.S. military is currently involved.

With the 2018 National Defense Strategy comes an opportunity for the Air Force to refocus its management of WRM. Centralized management with decentralized execution would support strategic resiliency and responsiveness goals.

---

[1] U.S. Department of Defense (DoD), *Summary of the 2018 National Defense Strategy of the United States of America: Sharpening the American Military's Competitive Edge*, Washington, D.C., 2018a.

Chapter Four

Tools to Enhance Air Force Management of WRM

Global managers need ways to validate warfighter demands, prioritize among theaters, optimize positioning, quantify and qualify the effect that different positioning strategies have on different AORs, convey analyses in terms of operationally relevant metrics, influence storage site selection, and report readiness. In this chapter, we suggest methods that a WRM global manager could employ to address these challenges.

In Chapter Two, we highlighted a portfolio of PAF and other RAND research that sought to address the challenges of WRM management. Most of the research focused on peacetime cost optimization, with some analyses looking at robust posture options that might enable more-effective operations across a range of military operations outlined by CCDRs. Drawing on prior research, and considering the challenges posed by the current national security environment, we focus this chapter on resilient and responsive management of WRM. We highlight concepts and tools that can enhance global management of WRM.

We start by addressing what joint policy highlights about the responsibilities of service WRM programs, as well as the dynamics of the planning environment within which those service WRM managers must work. We then describe concepts and tools that could be useful to a service's WRM manager trying to manage from a global perspective. These concepts and tools address requirements determination, management of WRM as capabilities versus individual items, considerations for developing posture options, decision aids to address logistics analysis factors, and tools to analyze posture options based on different objective functions (for example, operational effectiveness versus costs).

What Joint Policy Requires of a WRM Global Manager

Joint instructions provide guidance to the services about the management of their WRM prepositioning programs. Our analysis supports the position of Chairman of the Joint Chiefs of Staff Instruction (CJCSI) 4310.01D;[1] therefore, it is useful to take a moment to understand what that joint policy requires of an Air Force global WRM manager.

CJCSI 4310.01D starts by explaining the purpose of prepositioning WRM:

> [Prepositioned WRM (PWRM)] refers to war reserve materiel strategically located to facilitate a timely response in support of Combatant Commander (CCDR) requirements during initial phases of an operation. To respond quickly to requirements, Services shall size, manage, and pre-position resources in a

---

[1] Joint Chiefs of Staff, *Logistics Planning Guidance for Pre-Positioned War Reserve Materiel*, Chairman of the Joint Chiefs of Staff Instruction 4310.01D, Washington, D.C., December 30, 2016.

[geographic combatant command (GCC)] area of responsibility (AOR). Pre-positioning provides the greatest practicable flexibility to respond to a spectrum of regional contingencies while reducing the demand on the global transportation network.

Pre-positioned war reserve capabilities (PWRC) (ashore and afloat) place military unit type equipment and supplies at or near the point of planned use to reduce reaction time and ensure timely support of a specific force during the initial phases of an operation. These assets provide CCDRs with initial capabilities and materiel until establishment of a mature and sustained Strategic/Theater Distribution Network (TDP).

Services support GCC requirements by pre-positioning capabilities, forces and equipment, to support global requirements. Services take into account strategic direction, national security threats, strategic mobility, and GCC requirements to determine the size, composition, and positioning of PWRM.

The Services plan, program, budget and execute to the most demanding contingency plans to identify the pre-positioning requirements, derived from integrated planning and aligned with the top priority plans listed in the JSCP and Joint mobilization plans. PWRM capabilities and stocks enable prompt and sustained combat operations on air, land and/or at sea.[2]

These background paragraphs illuminate the challenges and trade-offs that a global WRM manager faces. The first paragraph of the excerpt brings to light the organization that the global manager should go to in order to understand the requirements—the combatant command. It also sheds light on the performance parameters or metrics by which WRM posturing decisions should be managed—that is, the "greatest practicable *flexibility* to respond to a *spectrum* of regional contingencies while *reducing* the demand on the global transportation network" (emphasis added). It calls for services to develop a robust WRM posture that reduces transportation demands.

The second paragraph in the excerpt bounds the services' problem a little more by suggesting that WRM should be "placed at or near the point of use to reduce reaction time and ensure timely support." So, yet another measure to be considered when developing a WRM prepositioning posture is the time required for resources to be in place to support initial operations. The second paragraph also calls out a key element of WRM management when it classifies the resources being positioned as "capabilities." This implies that the global WRM manager's role is about managing capabilities, not just items.

The third and fourth paragraphs help the services understand the dynamics they must consider when establishing and managing their prepositioned WRM programs. The CJCSI directs the services to align their programs with strategic guidance, work with CCDRs to understand the nature of the threat and the WRM requirements, and manage those capabilities based on priorities established by the Secretary of Defense.

---

[2] Joint Chiefs of Staff, 2016, p. A-1.

While the first excerpt highlights fundamental components of WRM management, such as metrics to consider, and frameworks for thinking about capabilities, a separate section provides specific elements for the services' WRM managers to consider, in coordination with the CCDR service components, when developing a plan for prepositioning those capabilities in forward locations:

> GCC Service components determine PWRM requirements—GCCs validate the requirement and the Services review the GCC's requirement to determine supportability based on current force structure and availability of materiel and equipment. The Services develop their PWRM strategy and plan to fill the GCC's requirement. However, supporting multiple [operations plans]/ contingency plans may require the same capabilities and stocks. Any identified shortfalls in PWRM requirements will include a mitigation plan to reduce the operational risk of the PWRM shortfall. To address shortfalls, Services/GCC(s) may submit issue nominations for inclusion into the Service Program Objective Memorandums.

> Services consider doctrine, organization, training, materiel, leadership and education, personnel, and facilities when determining the composition and positioning for pre-positioned materiel. Additionally, Services shall consider the following elements of logistics support analysis when identifying their PWRRs:

> - Access, basing, and overflight rights
> - Force requirements and sourcing
> - Verification of sustainment
> - Analysis of transportation
> - Assessment of networks and distribution network limitations
> - Logistics support refinement
> - Commodity distribution concepts
> - Force protection, both active and passive measures
> - Retrograde planning
> - Reset planning
> - Host Nation requirements
> - Time to deploy.[3]

These sections require the service's WRM manager to work closely with CCDRs to understand their requirements, recognizing that there likely will be a shortage of resources available to meet all of the CCDRs' requirements. Considering those shortfalls, service WRM managers must work with the CCDRs and their service components to understand the associated implications and risks. It is reasonable to assume that those risks should be communicated in operational terms consistent with the metrics highlighted in the first CJCSI excerpt (for example, the effect on strategic lift resources and the ability to respond quickly based on the CCDR's plan).

Finally, this second excerpt points to many logistics analysis factors that WRM managers should consider when developing their forward preposition posture. These factors speak to the

---

[3] Joint Chiefs of Staff, 2016, pp. A-2–A-3.

complexity of the planning environment within which WRM managers must work. Several factors are interrelated, and the decisions on posture often become trade-offs across many of these factors.

To further describe how managing WRM from a global perspective might look, we use the remainder of this chapter to explain several WRM-related concepts and tools developed by PAF researchers.

Requirements Analysis

One of the primary responsibilities of service WRM managers is to work with CCDRs and service components to understand the prepositioned WRM requirements for each theater. The Air Force has some tools that can validate warfighter demand for specific functions (for example, BEAR; munitions; and petroleum, oil, and lubricants [POL]) but not others. This section describes a tool developed by PAF researchers that produces a list of UTCs needed to support a specific operational scenario.[4] The tool can be used to validate warfighter requirements. Later in this section, we provide a concept for shifting from viewing WRM as items to viewing such resources as capabilities.

Lean-START

As described in Chapter Two, Lean-START is an Excel-based analytical tool for lean force package planning.[5] Lean-START generates equipment and personnel needs, as well as movement characteristics for associated transport, based on user-specified parameters. Planners can use it to generate requirements and consider the trade-offs between deployment speed, capability levels, and risks. Analysis in Lean-START is driven by the following:

- operational scenario: duration, type and amount of aircraft, mission type, and sortie rates (Figure 4.1)
- beddown planning: nonmaintenance agile combat support, showing the various options for the level of capability available by functional area (Figure 4.2)
- tailoring: user-provided input to adjust requirements based on existing levels of support from host nations, existing materiel, and so forth.

The outputs contain details about number of personnel, amount of equipment, and types of UTCs needed to support the inputs about scenario, beddown, and tailoring.

---

[4] UTCs are highly modular force package sets that specify the manpower and equipment required to fulfill a given capability.

[5] Mills et al., 2017.

Figure 4.1. Operational Scenario Inputs in the Lean-START Model

Duration		Weapon System & Flying Profile							
Total Duration (days)	10	Aircraft	Number	Mission	Sortie rate	ASD	Sorties/day	Personnel	Weight (Stons)
Max flying days	1	B-2			-		0		
Maintenance capability	Launch & recover	B-52			-		0		
		F-16CD	12	Air-to-Ground	1.50	1.00	18	111	66
		F-15C	0	No Munitions	-	-	0		
		F-15E	0	No Munitions	-	-	0		
		F-15J	0	No Munitions	-	-	0		
		FA-18	0	No Munitions	-	-	0		
		F-22	0	No Munitions	-	-	0		
		C-130	0		-		0		
		EC-130	0		-		0		
		MC-130	0		-		0		
		HC-130	0		-		0		
		AC-130	0		-		0		
		KC-130	0		-		0		
		KC-10	0		-		0		
		KC-135	0		-		0		

SOURCE: Mills et al., 2017, p. 27.

NOTE: ASD = average sortie duration.

Figure 4.1 shows a screen capture of the operational scenario inputs of Lean-START. The *total duration* of the operation drives care and feeding support requirements, such as billeting, food, and water. A sustained operation will have an extensive care and feeding footprint; a shorter one could have virtually no support. *Flying days* apply to aircraft operations and drive the maintenance footprint, munitions, and fuel. Maintenance *capability* includes levels of capability from basic flight-line servicing up to full-scale backshop repair.

Figure 4.2 shows the input categories and options for beddown planning. The selections below each heading show the choices that a user has for each category; in some cases, the user selects the known characteristics of the operating location, and in others, the user dials up and down the desired level of capability (for example, security forces and firefighting)

Lean-START can be used to assess trade-offs between the amount of personnel and equipment required to support an operational scenario, the amount of lift resources required to deploy the personnel and equipment, and how the movement requirements can be adjusted based on a commander's risk tolerance in different functional support categories. For example, if a commander decides to accept more risk in terms of fire and crash rescue capability and focus on protecting aircraft only as opposed to all structures at a forward location, Lean-START will show how that choice results in a reduction in requirement for lift.

Figure 4.2. Capability Options for Beddown Planning Inputs in the Lean-START Model

Base Input Parameter Options		
Base_Type	**Conventional Threat**	**CBRN Threat**
High Capability (MOB/COB)	High	High
Med. Capability (COB/IAP)	Low	Medium
Low Capability (COB)		Low
Austere	**Security_Forces**	
	Protect base	**EOD**
POL Supply	Protect airfield	Threat-driven
Austere	Protect flightline	Protect aircraft only
Storage/Fuel provided	SF Team	Minimal
Fuel + Fillstands provided	None	None
Full service provided		
	Firefighting	**CBRN Capability**
	Protect aircraft and structures	Threat-driven
POL Support	Protect aircraft only	Minimal
FORCE + fillstands	Extract pilot only	None
FORCE + trucks	C2 only	
ABFDS	No support	**Advanced ADR**
FARP		None
		Small
		Medium
		Large
		Very large

SOURCE: Mills et al., 2017, p. 28.

NOTE: ABFDS = aerial bulk fuel delivery system; C2 = command and control; COB = collocated operating base; EOD = explosive ordnance disposal; FARP = forward area refueling point; FORCE = fuels operational readiness capability equipment; IAP = international airport; MOB = main operating base; SF = security forces.

As one of its outputs, Lean-START generates a set of UTCs based on operational characteristics, beddown planning, and any tailoring input provided by the user. Based on the list of UTCs, Lean-START identifies which UTCs or parts of UTCs are good candidates for prepositioning based on category and equipment characteristics (for example, rolling stock versus palletized). In this analysis, the list of candidate WRM items served as the basis for other concepts and models that we applied in this chapter to demonstrate an approach for global WRM management.

To make the connection between desired operational capability and required personnel and equipment, Lean-START contains deployment logic or "rules" for which equipment and personnel deploy, under which circumstances (for example, number of aircraft, mission type, duration, and desired capability level). The rules in Lean-START were derived from official sources (for example, mission capability statements) and discussions with SMEs. The deployment logic contained in the tool represents a standardized, rule-based way for establishing deployment requirements and for informing decisions about what could reasonably be placed in WRM. Given information about the intended operating location's existing capabilities and capability selections in the tool, the deployment requirements can be appropriately assessed.

It is necessary to have rule-based models and decision-support tools so that requirements determinations and allocation decisions are standardized, repeatable, and transparent. This allows customers to have confidence in a globally managed process and to understand decision logic when their own demand goes unmet, and it provides a basis on which to advocate for resources or a higher priority. Without such a construct, customers may have little confidence in allocation decisions and may be less willing to subject their needs to those of the global enterprise. Lean-START can be used as a standard, repeatable process by a WRM global manager to validate warfighter demands.

Comparing Competing Theater Demands

As part of global WRM management, it is useful to understand how the capabilities required within WRM stocks vary by theater. Each region has different operational requirements, different types and numbers of prospective operating locations needed and available, different required closure times, and different transportation availability and constraints. These factors will, in turn, influence total demands for equipment, desired placement of WRM, and trade-offs and priorities among the regions.

To inform this discussion, we explored notional examples using USAFE and PACAF, two theaters currently developing their own WRM requirements and plans. The first step was to compare equipment requirements. We used Lean-START to generate requirements for USAFE and PACAF and then examined differences at the UTC level. We applied assumptions regarding the baseline capabilities of different locations within each theater, which affects the type and quantity of items needed from WRM. Because there was a slight difference in the total number of base locations per theater (28 in USAFE and 27 in PACAF), we compared the average number of UTCs per location instead of the total number of UTCs per theater. Thus, our analysis does not address variations at the individual base level.

We focused the Lean-START output on items with a non-zero WRM weight. USAFE and PACAF had an average of 173 and 103 UTCs per location, respectively. In USAFE, 38 percent of the average number of UTCs were in common with those in PACAF, whereas 64 percent of the average number of UTCs in the PACAF requirements list were common to both. The average weight was also higher in USAFE—1,477 tons per USAFE location versus 991 tons per PACAF location. However, the average weight per UTC was slightly higher in PACAF than USAFE, at 9.7 tons per PACAF UTC versus 8.5 tons per USAFE UTC. The differences in the average number and weight of UTCs by location suggest that varying assumptions regarding baseline location capabilities, including host-nation contributions, affects the total amount and weight of WRM assets.

We conducted similar analysis of the Air Force–generated WRM packages commonly referred to as DABS. The DABS kits for USAFE had 79 unique UTCs, whereas PACAF DABS kits had 62 unique UTCs. Overall, there were 20 UTCs in common across both theaters.

The Lean-START and DABS comparisons are examples of the types of analysis that a global WRM manager should be conducting. A global WRM manager might want to better understand how a functional area (for example, civil engineering) is approaching the beddown challenge for each theater. Why are there differences? What drives the differences? Are there processes in one theater than can be applied globally, thus making WRM capabilities interchangeable across AORs?

Drawing comparisons between the DABS packages for PACAF and USAFE is complicated by the way the Air Force thinks about UTCs and how the DABS packages are constructed. The 2015 version of AFI 25-101 defines WRM as "packaged capability Unit Type Codes (UTC). These capability packages are composed of equipment, vehicles, consumables, munitions, and medical resources."[6] The same policy document directs the Air Force Global WRM Management Office to identify CCMD "requirements by UTC (capability) and required prepositioning location."[7] Thinking of WRM capabilities in these terms presents challenges for a global manager.

Many UTCs in the Air Force consist of a single item, vehicle, or piece of equipment—or, in some cases, a single airman with specific skills. This approach to UTC design and management is useful from a functional perspective when an Air Force functional manager knows exactly what a particular vehicle or piece of equipment can do. The challenge is whether these UTCs describe a capability that is meaningful to a CCDR's or service component's planning staff as they are building a force list for a beddown location. In addressing this for our global WRM management example, we developed an alternative approach to how the Air Force might think of managing WRM, as capabilities rather than individual resources.

Capabilities Versus Items

Considering CJCSI 4310.01D, on prepositioned WRM capabilities, we developed a construct for thinking about WRM in capability buckets that might be meaningful to CCMD planners as they plan how to deploy resources from WRM storage locations at the commencement of conflict. We considered several elements when thinking about what items or resources would fall into each bucket.

First, given that WRM is designed to enable forward locations to more quickly receive the *capabilities* needed for initial operations, we recognized the temporal aspect of the capabilities and focused on what capabilities were needed in what order. In doing this, we departed from some functional area deployment policies in order to minimize the transportation time for the most-urgent capabilities. This was evident when looking at such resources as BEAR, for which we assumed that, under the most time-critical situations, a commander might be willing to accept a reduced level of base operating support (BOS) for a period of time (perhaps 30 days) in order to increase the number of forward locations that can be open over a specified time frame.

We also created broad buckets that align with which activities airmen might want to perform at a base at different times. The categories are designed to correlate to basic mission activities

---

[6] Department of the Air Force, 2015, p. 6.

[7] Department of the Air Force, 2015, p. 15.

associated with opening a FOL, commencing sortie generation, and supporting a base population for a short period of time and then a longer period of time. Finally, given that our analysis is based on operating in a contested environment, we recognized the need for a capability to recover aircraft and the base after an attack. We bucketed WRM capabilities into the following categories, which are ordered based on how quickly each is needed (*temporal order*):

- *Category 1: Open the base and prepare to receive forces*; example capabilities include material handling equipment, force protection, and communications
- *Category 2: Force projection*; example capabilities include maintenance, sortie generation, fuels, munitions, crash rescue, and explosive ordnance disposal
- *Category 3: BOS initial operating capability (IOC)*; example capabilities include minimal housekeeping, a special operations–like capability, power production, and limited industrial BEAR
- *Category 4: Recover the base and aircraft*; example capabilities include rapid airfield damage recovery and aircraft battle damage repair
- *Category 5: BOS full operating capability (FOC)*; example capabilities include full housekeeping and general-purpose vehicles.

The result of the Lean-START, DABS, and capability bucket analysis provided us with a general view of the types of capabilities that should be kept in prepositioned storage sites. Having defined the buckets of WRM by category and put them in temporal order, we turned our attention to posturing WRM capabilities in different locations based on the capability bucket in which they fall. We developed a decision tree that yields posture alternatives for items in each category, discussed in the next section.

Creating WRM Posture Options

We developed a decision tree framework to connect different levels of critical capabilities to a flexible array of posture options. The decision tree considers the timing of need (which relates to the temporal order of the five categories), transportability, and the cost of the asset. Based on these factors, the decision tree leads to one of four posture options, ranging from centralized (in CONUS or in theater) to more-distributed postures (moderately distributed or at point of use). A more centralized posture represents items that are not needed immediately, are easily transportable, and are more expensive. Items difficult to transport (oversized or heavy), items available in the local economy, and items needed quickly are pushed toward more-distributed postures (see Figure 4.3). When appropriate, the decision tree incorporates the flexibility to select two posture options per asset.

30

Figure 4.3. A Notional Framework for Thinking About WRM Postures

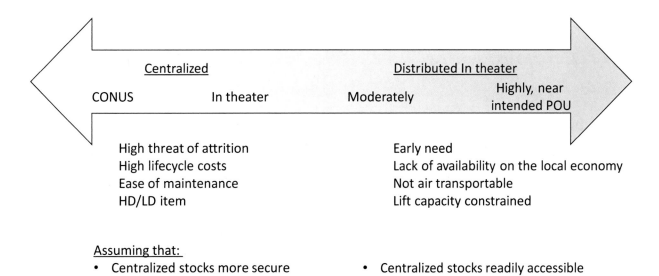

NOTE: HD/LD = high demand/low density; POU = point of use.

A full overview of the decision tree framework is shown in Figure 4.4. The first step considers the ranking of individual capabilities by temporal order—that is, how quickly those resources are needed. Categories 1 and 2 (outlined earlier) are grouped together, while categories 3 through 5 are considered as separate branches of the decision tree. The next step considers the transportability of the asset. For each UTC, we define an easily transportable asset as one that both is transportable by air and is not outsized.[8] If either of these conditions is false, then the asset is considered not easily transportable. The next branch of the decision tree incorporates the cost of the asset, which could affect the decision to buy additional stocks and distribute widely. There will be some cost threshold that makes an item so expensive that it becomes precious. We do not recommend a particular value in our analysis; however, we use $50,000 as a notional cost threshold in this example.[9] If the asset costs more than $50,000, then the asset is put in a more centralized location; if the cost is less than this cost threshold, then the asset is shifted toward more-distributed postures.

---

[8] In this example decision tree, transportability captures an asset's size and whether it can be transported by air. Outsized cargo—that is, any single piece of cargo that exceeds any one dimension (1,000 in. long, 117 in. wide, and 105 in. tall, for example)—cannot be transported by civilian contract cargo carriers. Another level of detail, such as knowing how many C-17 sorties would be needed to deploy the asset, could further improve decisionmaking; however, we did not evaluate that detail in this analysis.

[9] We used $50,000 as a notional threshold to capture some vehicles and rolling stock but to eliminate assets that are highly technological (and might need to be updated), require additional maintenance, or require extra considerations for storage. Global managers could decide that certain assets that are higher value (for example, fire trucks) should be considered WRM and stored accordingly or that another monetary value is more suitable for the analysis.

Figure 4.4. Overview of the Decision Tree Framework Used to Provide Posture Options

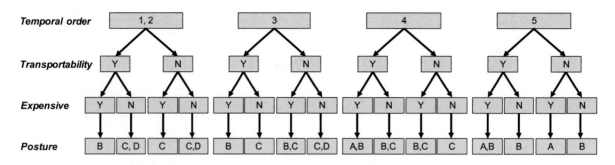

Key

Mission Criticality
1. Prepare to receive forces
2. Force projection
3. BOS IOC
4. Recover the base / aircraft
5. BOS FOC

Posture
A. Centralized in nearby theater or CONUS
B. Centralized in theater
C. Moderately distributed in theater
D. Highly distributed in theater

In Figure 4.5, we show the application of the decision tree framework to two UTCs with different temporal order. The first example is an all-terrain forklift, which is defined as temporal order (or category) 1 because it is needed to prepare the base to receive forces. Although this asset can be transported by air, it is outsized, so it is not considered easily transportable. The asset is also expensive, so the decision tree suggests a posture that is moderately distributed in theater. If the forklift were less expensive, it would be more highly distributed, and if it were easier to transport, it would be shifted toward a centralized posture. In contrast, Figure 4.5's second example is a laundry unit, which falls under temporal order 5 (BOS FOC). This asset is considered easily transportable because it can be transported by air and is not outsized. Because the item is expensive, the decision tree suggests more-centralized postures, either centralized in nearby theater or CONUS or centralized in theater. If the laundry unit were not expensive, it would have been pushed toward posturing it centrally within the theater.

Figure 4.5. Decision Tree Analysis Example for Two Types of WRM

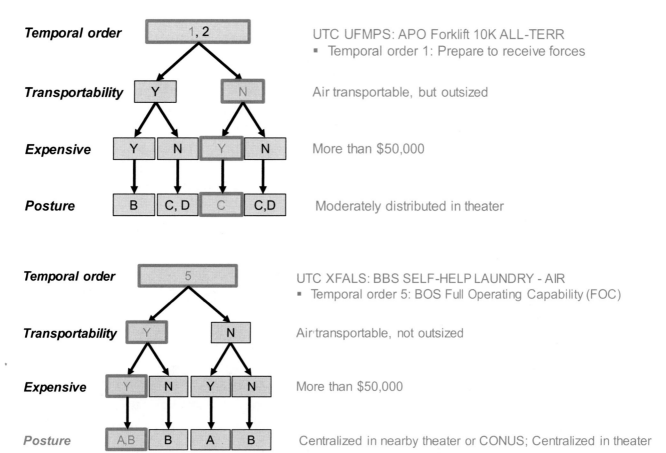

There are several flexible assumptions built into the decision tree framework that could be tailored by a global WRM manager. In the illustrative decision tree described here, we selected initial values for the transportability and cost of an asset that could be modified. First, we defined the transportability of an asset to depend on its ability to be transported by air and to not have any component that is considered outsized, but those parameters might not always be appropriate. Second, we defined expensive items as those costing more than $50,000, but this level could be adjusted as needed. In addition, the decision tree framework does not consider whether an asset is available in the local economy or through host-nation support. The global WRM manager could choose to exclude certain assets from posture considerations and instead rely on locally acquiring those capabilities.

Analyzing Various WRM Preposition Posture Options from a Global Perspective

Managing WRM from a global perspective drives a level of analysis that enables the service WRM manager to answer several questions. As we highlighted earlier in this chapter, the WRM manager must be capable of communicating the risks and implications with various postures, assuming that resources will be constrained. In the context of the 2018 National Defense Strategy and the international security environment characterized by adversaries capable of contesting the Air Force's freedom of movement, there are several questions the posture analysis should answer, such as the following,

- Given limited resources to purchase additional WRM to preposition, which postures provide the most robust ability to respond to a range of potential contingencies?
- Given limited strategic transportation resources to move prepositioned stocks to their intended points of use, which postures enable the greatest number of available sites in the shortest amount of time? At what cost?
- Given some prioritization of one CCDR's capability demands over another, what postures best mitigate the risks of responding in a timely fashion to lower-priority requirements?

In this section, we use a PAF-developed model to demonstrate the type of analysis that a global manager might need to conduct in order to answer these questions. Elements of this analysis address the logistics factors that CJCSI 4310.01D suggests should be addressed during the course of developing a prepositioning strategy.

The PRePO Model

The Prepositioning Requirements Planning Optimization (PRePO) model is a mixed integer linear program optimization model designed to answer strategic-level questions about prepositioning WRM in theater either for a specific regional conflict or globally to meet multiple regional conflicts.[10] The PRePO model can be used to find performance-optimal WRM prepositioning postures that enable transporting WRM from storage locations to end-use locations in the shortest time possible.[11] It can also be used to assess user-developed postures to determine if the timeline to transport the required WRM to the base is achievable, given such inputs as WRM requirements at FOLs, throughput capacities at facilities, and numbers of transportation assets available (see Figure 4.6). The model uses a mixed integer linear program solver to find optimal solutions and then outputs storage allocations among candidate sites; costs of storage and transportation; and a transportation schedule of cargo broken down by storage site, operating location, transportation asset, and cargo type.

---

[10] The PRePO model and its application in previous analyses are discussed in detail in a previous document. See Brent Thomas, Bradley DeBlois, Katherine C. Hastings, Beth Grill, Anthony DeCicco, Sarah A. Nowak, and John A. Hamm, *Developing a Global Posture for Air Force Expeditionary Medical Support*, Santa Monica, Calif.: RAND Corporation, forthcoming, Not available to the general public.

[11] An optimal allocation using the PRePO model depends on the objective function used. Past analyses have used both the lowest cost and the fastest time to get the required WRM to the base as objective functions. For this analysis, *optimal* strictly means the allocation that provides the required WRM in the fastest time, in order of WRM priority category. This is discussed in more depth in Appendix C.

Figure 4.6. Inputs and Outputs of the PRePO Model

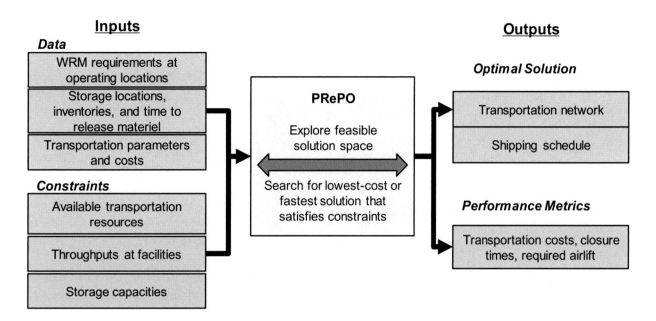

There are some differences in the PRePO model used for this analysis and the one used in previous RAND analyses.[12] To present analyses relevant to a global WRM manager, we made changes to enable the model to prioritize the transportation of some WRM types before others and to some locations before others. In short, the changes enable the model to consider timing differences across WRM types, bases, and theaters. The timing capability enables the model to make meaningful trade-offs between positioning WRM close to the point of intended use and centralizing it at a more distant location.

PRePO Analysis

We assessed the cost and global performance of a set of WRM posturing COAs (posture plus availability of transportation) for conflicts in four global regions (referred to here as regions A through D). Consistent with 2018 National Defense Strategy guidance, only two of those regions contained prepositioned WRM. These analyses demonstrate the sort of capability that could help a global WRM manager make trade-off decisions about cost, risk, and performance on a global scale.

The results presented in this section show the maximum number of bases that could receive the WRM required for IOC over time, once there has been some strategic warning of an imminent conflict (the warning takes place on day zero) and the process of transporting WRM to operating locations has begun. In this report, we sometimes use the terms *closure* and *closed* to mean that a base has received either its full WRM requirement or its total requirement for a specific WRM type.

---

[12] Thomas et al., forthcoming.

The results in the following figures assume the best possible use of the transportation assets available for a given COA—when the highest priority is given to bases receiving the WRM required for IOC. The COAs shown in the results include one in which WRM is moderately distributed and one in which WRM is centralized regionally. Each COA also indicates that the availability of transportation assets is either low or high.[13] Figure 4.7 shows the number of bases in region A that have received the WRM required for IOC between days 0 and 30 when a centralized regional posture (with WRM storage at two locations) is used and the availability of transportation assets is low. The figure shows that, at day 0, one base has already received all of its WRM required for IOC because, in the centralized storage posture in this region, WRM is stored collocated with one of the FOLs. At day 6, another base has received all of its WRM required for IOC. As time progresses, more and more bases receive the WRM required for IOC until, at day 17, all bases have received their full requirement.

Figure 4.7. Region A Bases Receiving the WRM Required for IOC over Time: WRM Storage Centralized Regionally, Low Number of Transportation Assets

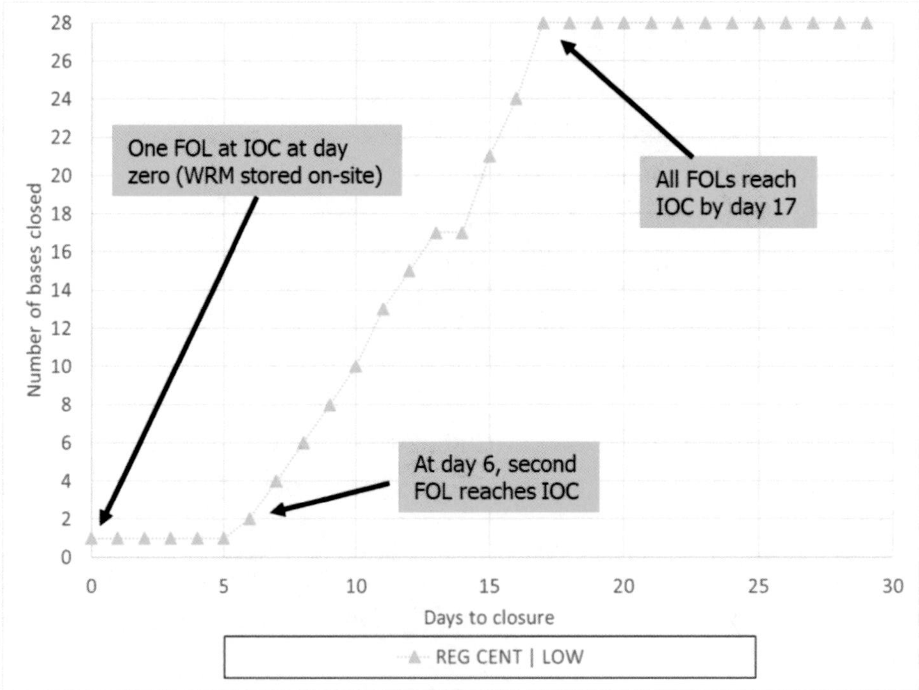

NOTE: REG CENT = storage is centralized regionally.

Figure 4.8 shows the improvement possible in the number of bases whose requirements are closed (that is, the bases received the WRM required for IOC) when the availability of transportation assets is high. Because both solutions (high and low levels of transportation assets)

---

[13] More details about the PRePO model, the storage COAs, and transportation assumptions are provided in Appendix C.

start from an identical storage posture, they both start with one base having received all WRM required for IOC. However, greater availability of transportation assets enables an additional base to receive its WRM required for IOC four days earlier (on day 2 rather than day 6); by day 8, the higher transportation level has allowed for nine more bases' IOC WRM requirements to be closed. Under the higher transportation assumption, all bases received the WRM required for IOC by day 13—four days earlier than when fewer transportation assets were used.

Figure 4.8. Region A Bases Receiving the WRM Required for IOC over Time: WRM Storage Centralized Regionally, High Versus Low Number of Transportation Assets

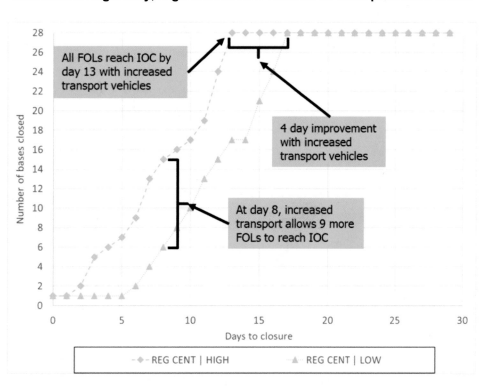

Figure 4.9 shows the effects of increasing the dispersal of WRM storage rather than increasing the amount of transportation assets available. The figure compares the COA from Figure 4.7 with a moderately distributed storage posture, in which WRM is prepositioned at six storage sites collocated with bases and is distributed throughout the theater. Under this posture, at day 0, six bases already have the WRM required for IOC because of collocated storage sites. Although WRM is more distributed under the storage posture depicted in blue, both of the COAs have the same level of transportation assets available, and bases' WRM requirements for IOC close as quickly as possible. By day 13, the bases with the more dispersed posture have received all of the WRM required for IOC, which is the same time frame as the centralized posture with more lift available (Figure 4.8) and four days earlier than the centralized posture with the same lift available.

Figure 4.9. Region A Bases Receiving the WRM Required for IOC over Time: WRM Distributed Across the Region Versus Centralized Regionally, Both with Low Number of Transportation Assets

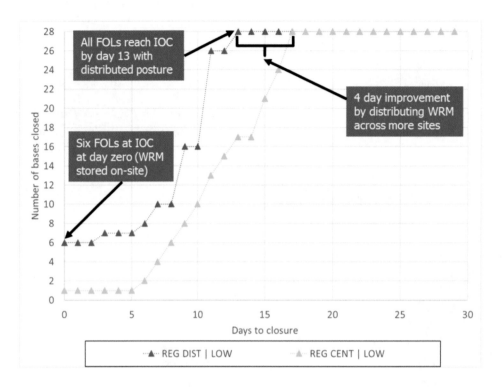

NOTE: REG DIST = storage is moderately distributed across the region.

Figure 4.10 compares the number of bases receiving the WRM required for IOC over time using the moderately distributed storage posture with low and high availability of transportation assets. Both COAs start out with six bases' WRM requirements closed. However, the greater available lift allows one of the COAs to start closing the bases' requirements much earlier and at a faster rate. The additional storage sites enable more cargo throughput, which allows more cargo to be transported during a single day (relative to the centralized postures from Figures 4.7 through 4.9), and the additional lift assets are able to take advantage of that increased throughput potential.

Figure 4.10. Region A Bases Receiving the WRM Required for IOC over Time: WRM Distributed Across the Region, High Versus Low Number of Transportation Assets

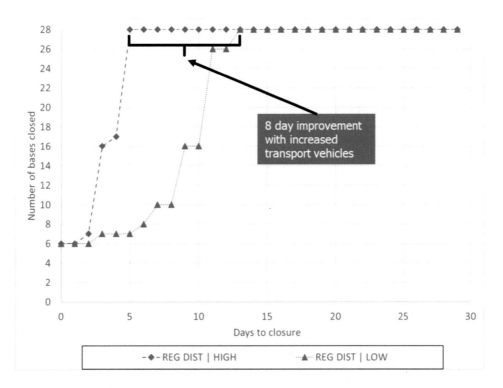

Figure 4.11 shows the differences in bases receiving the WRM required for IOC over time using the regionally centralized storage posture with high numbers of lift assets available versus the moderately distributed posture with low numbers of lift assets available. As the figure shows, the more distributed posture starts out with six bases' requirements closed at day 0, but the number progresses at a slower rate than the centralized posture with less lift available. Having a greater amount of lift available enables the centralized posture to catch up with the more distributed posture by day 5. Results using the two approaches are comparable after day 5, and both approaches see all base requirements closed by day 13. Clearly, multiple strategies can be used to achieve a desired result. Analysis of this sort can quantify the trade-offs for cost, performance, and relative risk for each posture.

Figure 4.11. Region A Bases Receiving the WRM Required for IOC over Time: WRM Storage Centralized Regionally with High Number of Transportation Versus WRM Distributed Across the Region with Low Number of Transportation Assets

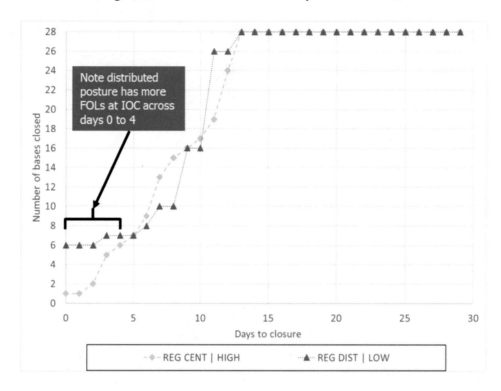

The next two figures again show the number of bases that receive the WRM required for IOC over time, but these figures depict region C, which faces a different conflict. Region C is one of the regions that does not contain prepositioned WRM, so it must rely on WRM that is stored in other regions. The results shown in Figure 4.12 represent centralized storage in the regions that do contain prepositioned WRM, along with an assumption of low numbers of lift assets available. Accordingly, on day 0, no bases have the WRM required for IOC. However, on day 1, region C can close a single base's requirements by strategic airlift alone. Note that regional transportation assets cannot be used until IOC WRM is repositioned by ship to a regional port, at which point trucks and regional airlift and sealift can begin to move the materiel. This happens by day 13, when a large ship arrives at port in region C, and all of the regional assets immediately come into play and begin transporting WRM to operating locations. Regional and strategic lift continue to contribute to closing base requirements until day 23, at which point all IOC-required WRM has arrived at the bases.

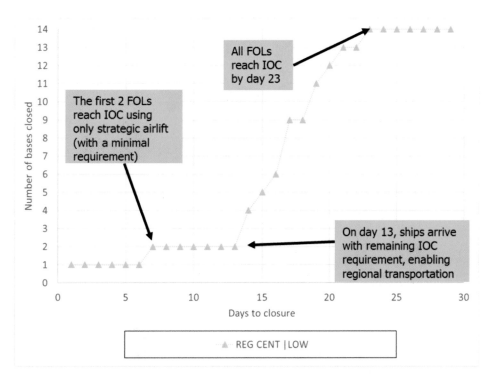

In the results from Figure 4.12, the primary driver of delay in getting IOC WRM to bases is the time it takes to transport the resources from storage locations in other regions to ports, then load ships, and then move ships by sea to ports in region C. There are several ways to reduce this time. One is by using distributed regional storage, because additional storage locations enable more trucks per day to move WRM to ports, and ships can load faster. However, this option can provide only limited improvement. There are only so many transportation assets available, so the full throughput capability of using multiple storage sites cannot be leveraged completely. Increasing the total number of lift assets is another way to speed up the loading of ships, given a distributed storage posture. However, the improvement here is also limited because there is a limit to how quickly ships can load and be ready for transport. Increasing lift also means that WRM can be transported faster regionally, to the extent that ships can unload quickly enough to take advantage of the increased trucking available.

In Figure 4.13, the blue line represents the improvement possible by using regionally distributed storage and increasing the total lift available. Under this COA, more strategic airlift is available to move WRM across regions, and ships are able to reposition WRM to regional ports sooner (by two days). Likewise, regional lift assets are able to move repositioned WRM sooner to close the last base's IOC WRM requirements four days sooner than when WRM is centralized (in other regions) and fewer lift assets are available.

Distributing storage and using more lift helps the cargo ships leave sooner, but these strategies cannot affect the biggest driver of the delay, which is the time it takes ships to travel from the region where WRM is stored to local ports in the region demanding WRM. The only way to decrease that shipping time is to move WRM storage closer to the region where it is needed. The red line in Figure 4.13, depicting the globally distributed option in the figure, reflects this option: Two of the WRM storage sites are repositioned within region A to a site on the boundary between regions A and C (but still inside region A).

Figure 4.13. Region C Bases Receiving the WRM Required for IOC over Time: WRM Stored Globally with Low Number of Transportation Assets Versus Distributed Across the Region with High Number of Transportation Assets Versus Centralized Regionally with Low Number of Transportation Assets

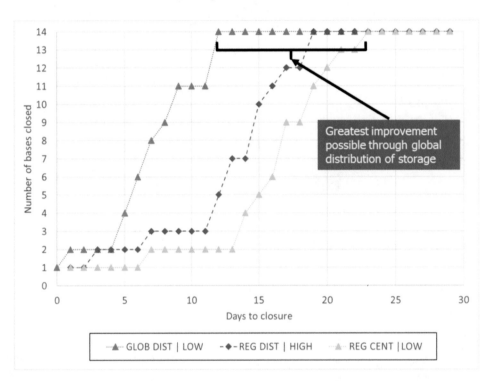

NOTE: GLOB DIST = storage is distributed globally.

As Figure 4.13 shows, moving storage sites closer to AOR boundaries, and thus decreasing the amount of time required for ships to travel between storage sites and operating locations, significantly improves the timelines for closing requirements, even without increasing the amount of lift available. Because the storage sites are significantly closer, strategic airlift is more effective and regional lift assets can get involved much sooner. The globally distributed option can close all region C bases' WRM requirements for IOC by day 12, a further improvement of seven days over the figure's blue line representing distributed storage and a high level of transportation assets.

However, it is important to consider that repositioning WRM in region A to sites near the border between regions A and C potentially means moving WRM farther from FOLs in region A. In the globally distributed posture examined, two of the six storage sites in region A are so

42

repositioned, at some cost to the speed with which region A IOC base requirements close. Figure 4.14 shows the effect of repositioning two of these storage sites (shown as the globally distributed, low transportation COA, in red). As the figure shows, the globally distributed posture starts on day zero with fewer bases having WRM requirements closed and finishes closing the region A bases nine days after the moderately distributed COA, which has all six bases located centrally in the region, irrespective of the needs of other regions. However, most of this delay in the globally distributed COA is caused by a few bases that are difficult to close; by day 9, this COA shows 19 of the region's bases (around 70 percent of the bases in the beddown) having received the WRM required for IOC, while the other two COAs have only 16 bases' IOC requirements closed by day 9.

Figure 4.14. Region A Bases Receiving the WRM Required for IOC over Time: WRM Distributed Across the Region with Low Number of Transportation Assets Versus Centralized Regionally with High Number of Transportation Assets Versus Globally Distributed with Low Number of Transportation Assets

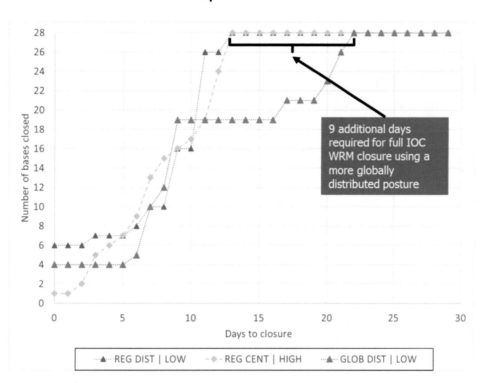

As Figure 4.14 shows, the model can quantify the effect of global prepositioning choices for decisionmakers. If only 70 percent of bases in region A are needed quickly, perhaps because not all of the required forces can mobilize and deploy within that time, then perhaps the globally distributed prepositioning scheme is the best option, with minimal additional risk. If this globally distributed posture increases risk too much with respect to region A, repositioning only one storage site (instead of two) may provide a more attractive solution. This type of modeling could be invaluable to the global WRM manager for quantifying the effect of prepositioning options.

Metrics and Reporting

One final area that we raise for consideration in WRM management is readiness reporting for mission support capabilities. Today, unit-level mission generation capabilities can be translated into capacity assessments for operationally relevant metrics—for example, the ability to support an operations plan.[14] The same is not true for mission support capabilities. Although the 635 SCOW does track and report WRM status for the commodities it manages, even in that small sampling, readiness reporting on WRM equipment is inconsistent across commands and may not reflect the actual condition of the equipment. A recent evaluation found that gaps in policy and information systems might be part of the problem.[15]

WRM managers need consistent measures of WRM readiness to conduct global assessments. Improving the consistency and quality of WRM equipment readiness reporting would help, but such data currently exist as inventories of individual items, not in terms of the capabilities the resources support. If the Air Force were to bin WRM equipment into aggregate operational metrics—for example, number of operating locations that can be supported[16]—both COMAFFOR planners and global managers would benefit. Global managers could assess available capability and predicted shortfalls. COMAFFOR planners could compare available capabilities to combatant command plans and then replan if shortfalls exist. The Air Force could choose to track these mission support readiness metrics in the Status of Resources and Training System (SORTS) or the Defense Readiness Reporting System (DRRS). Or it could expand the SharePoint site that the 635 SCOW already maintains. The key is to translate inventories into metrics that measure mission support capabilities.

---

[14] Anthony D. Rosello, Muharrem Mane, Jeffrey R. Brown, Emmi Yonekura, Henry Hargrove, Alexander Halman, Paul W. Mayberry, Dan Madden, Nahom M. Beyene, and C. R. Anderegg, *Air Force Readiness Reporting Relation to Operations Tempo and Suggested Improvements*, Santa Monica, Calif.: RAND Corporation, 2018, Not available to the general public.

[15] In 2018, the Air Force established a Sustainment Review Team to evaluate readiness of Air Force parts, supply, and equipment. The team identified several issues, including one about the reporting policy for WRM equipment readiness and information systems (the report is not available to the general public).

[16] Patrick Mills, John G. Drew, John A. Ausink, Daniel M. Romano, and Rachel Costello, *Balancing Agile Combat Support Manpower to Better Meet the Future Security Environment*, Santa Monica, Calif.: RAND Corporation, RR-337-AF, 2014.

Chapter Five
Strategic and Political Risk Assessment Tool

Several additional political, economic, and strategic factors beyond relative costs and transportation timelines can affect the viability of WRM prepositioning postures. Such factors include a host nation's long-term relationship with the United States and its internal political and economic stability. However, given that many of these considerations defy easy quantification because of their qualitative nature, researchers and practitioners have thus far incorporated these factors into planning only idiosyncratically or through background assumptions that bound the problem under study rather than as an integral part of the analysis of candidate site locations for WRM prepositioning.

To supplement the PRePO analysis from Chapter Four, we developed a diagnostic framework that can help logistics planners systematically assess a range of qualitative political and strategic risk factors that may affect access to and availability of prepositioned WRM assets. SPRAT builds on previous RAND work evaluating potential partner-nation support, including the RAND Security Cooperation Prioritization and Propensity Matching Tool, a diagnostic tool built in Microsoft Excel that helps decisionmakers identify mismatches between the importance of a country to U.S. interests, U.S. security cooperation funding to that country, and the propensity for successful U.S. security cooperation with that country.[1] SPRAT derives measures and underlying data from the Propensity Matching Tool, as well as other applicable sources (as described in Appendix D), to provide a succinct and systematic summary of more than a dozen characteristics across 195 countries that can assist planners in evaluating each country's suitability for WRM prepositioning.[2] Capturing a range of factors that could influence a potential host nation's willingness and ability to store and provide access to prepositioned WRM assets, SPRAT draws almost entirely on open-source data and information sources to code and categorize these factors in an accessible, user-friendly Microsoft Excel spreadsheet appropriate for A4 planners. The goal was to develop a rough snapshot of the kinds of considerations a planner might want to consider when developing prepositioning postures. SPRAT is not meant to provide a complete analysis of the potential for security cooperation arrangements between the

---

[1] See Christopher Paul, Michael Nixon, Heather Peterson, Beth Grill, and Jessica Yeats, *The RAND Security Cooperation Prioritization and Propensity Matching Tool*, Santa Monica, Calif.: RAND Corporation, TL-112-OSD, 2013, pp. ix–xii.

[2] SPRAT uses some of the same information as the Security Cooperation Prioritization and Propensity Matching Tool, simply updating the data or using the measures that the Propensity Matching Tool aggregates into a high-level measure. Other measures were derived from other RAND analyses, such as Stacie L. Pettyjohn and Jennifer Kavanagh, *Access Granted: Political Challenges to the U.S. Overseas Military Presence, 1945–2014*, Santa Monica, Calif.: RAND Corporation, RR-1339-AF, 2016; and Jennifer Kavanagh, *U.S. Security Agreements in Force Since 1955: Introducing a New Database*, Santa Monica, Calif.: RAND Corporation, RR-736-AF, 2014.

United States and other countries identified in this analysis. It is meant to demonstrate a preliminary screening tool that can be used by planners when considering positioning WRM assets. A more complete framework for systematically assessing partner-nation capabilities can be found in recent RAND work on security cooperation and capacity-building efforts.[3]

The following sections discuss the development of this tool and provide an overview of the measures that serve as proxies for different characteristics affecting the willingness, reliability, and capability of potential host nations to preposition WRM assets. We then provide an illustration of how the information supplied by SPRAT can supplement optimization modeling to provide an initial assessment of candidate site locations for WRM prepositioning.

SPRAT Motivation and Methodology

Political and strategic factors beyond those already incorporated into previous quantitative tools and optimization modeling affect the viability and reliability of WRM prepositioning postures. Although such considerations may prove immaterial when considering the prepositioning of WRM assets with reliable partners, the risk that a candidate host nation would deny access to prepositioned assets for political or strategic reasons may outweigh any benefits that would otherwise accrue from staging these assets in cost-optimized locations. For example, Turkey's last-minute refusal to allow U.S. forces access to its territory and airspace in advance of Operation Iraqi Freedom underscores that neglecting political factors can erode strategic and operational effectiveness.[4] Overlooking these considerations when evaluating the costs and benefits of various prepositioning postures can have significant ramifications should access to WRM storage sites be denied during contingency operations.

To identify and operationalize these factors, we conducted a wide-ranging literature review across different fields of study and consulted with SMEs who had experience in building similar diagnostic tools or who had conducted previous analyses of prepositioning logistics in limited and denied operational environments. Guiding this research were two key questions:

- What political, economic, and strategic factors might affect the willingness of host nations to grant the United States access to prepositioned WRM assets within their borders during peacetime and contingency operations?

---

[3] See David E. Thaler, Beth Grill, Jefferson P. Marquis, Jennifer D. P. Moroney, Heather Peterson, Lisa Saum-Manning, and Ilana Blum, *Assessing Partner-Nation Air, Space, and Cyber Capabilities: Supporting Development of Security Cooperation Strategy for the U.S. Air Force Flight Plan*, Santa Monica, Calif.: RAND Corporation, 2018, Not available to the general public.

[4] The refusal of Turkey's parliament to authorize access to U.S. forces required the United States to heavily revise its operations plans. For the consequences of Turkey's denial of access on Air Force support planning during Operation Iraqi Freedom, see Kristin F. Lynch, John G. Drew, Robert S. Tripp, and C. Robert Roll, Jr., *Supporting Air and Space Expeditionary Operations: Lessons from Operation Iraqi Freedom*, Santa Monica, Calif.: RAND Corporation, MG-193-AF, 2005, pp. 48–56.

- What political, economic, and strategic factors might affect the capacity of host nations to reliably secure and make available prepositioned WRM assets during peacetime and contingency operations?

We drew from research on political risk assessment in the logistics and supply chain literature that examines what factors businesses, disaster relief organizations, and other nongovernmental entities consider when deciding whether to invest in foreign markets and where to locate their facilities and supply inventories aboard. According to its survey of global investors, the World Bank's Multilateral Investment Guarantee Agency reported in 2013 that political risk perceptions ranked at or near the top of considerations that investors weigh when operating in developing markets.[5] Similarly, when prepositioning their inventories, humanitarian relief organizations weigh such factors as the availability and quality of labor, business support services, infrastructure, government stability, administrative transparency, corruption levels, and attitudes toward humanitarian assistance alongside efficient cost and proximity concerns.[6] When making plant location decisions, multinational corporations weigh similar considerations; they also tend to examine the location decisions of their competitors, either to compete directly for market share or a similar pool of low-cost laborers or high-quality talent (as is the case in Silicon Valley) or to take advantage of an untapped market.[7]

We also consulted SMEs to help identify relevant metrics (see Table 5.1 later in this chapter). As an early source of inspiration for the tool's design and content, we studied the RAND Security Cooperation Prioritization and Propensity Matching Tool, described earlier.[8]

The Propensity Matching Tool assigns each of 195 countries an overall security cooperation propensity score, which aggregates sub-scores across ten categories comprising 66 measures and summarizing relevant characteristics of potential partner nations, such as the strength of the nation's economy and governance institutions; its support for its military; its vulnerability to internal and external security threats; and the nature of its long-term political and economic

---

[5] Political risk ranked at the top of concerns among investors surveyed in four of the past five years that the survey was conducted. See Multilateral Investment Guarantee Agency, *World Investment and Political Risk 2013*, Washington, D.C.: International Bank for Reconstruction and Development/World Bank, December 2013. For an empirical application, see James B. Mshelia and John R. Anchor, "Political Risk Assessment by Multinational Corporations in African Markets: A Nigerian Perspective," *Thunderbird International Business Review*, February 2018.

[6] Deila A. Richardson, Sander de Leeuw, and Wout Dullaert, "Factors Affecting Global Inventory Prepositioning Locations in Humanitarian Operations—A Delphi Study," *Journal of Business Logistics*, Vol. 37, No. 1, 2016, p. 61. See also Thomas E. Lang and Ronald G. McGarvey, "Determining Reliable Networks of Prepositioning Materiel Warehouses for Public-Sector Rapid Response Supplies," *Advances in Operations Research*, 2016.

[7] Alan David MacCormack, Lawrence James Newman III, and Donald B. Rosenfield, "The New Dynamics of Global Manufacturing Site Location," *Sloan Management Review*, Vol. 35, No. 4, Summer 1994. For an empirical analysis, see Rohit Bhatnagar, Jayanth Jayaram, and Yue Cheng Phua, "Relative Importance of Plant Location Factors: A Cross National Comparison Between Singapore and Malaysia," *Journal of Business Logistics*, Vol. 24, No. 1, 2003.

[8] Paul et al., 2013.

relationship with the United States, including its track record with U.S. security assistance and foreign aid.[9]

The findings of several other previous RAND studies informed our thinking about which characteristics would be important to include in our framework. Work on the U.S. overseas military presence led us to consider how a country's previous experience with hosting U.S. forces and maintaining a permanent U.S. basing presence should factor into prepositioning planning.[10] In particular, Stacie Pettyjohn and Jennifer Kavanagh's comprehensive analysis of political challenges to U.S. overseas military access during peacetime and contingency operations between 1945 and 2014 provided important insights into how a host nation's prior history of access restrictions, denials, and evictions could shape expectations for denial of access to prepositioned WRM assets.[11]

SPRAT Categories, Measures, and Data Sources

At the conclusion of our consultations and literature review, we identified 15 measures that were organized into two broad categories designed to capture the factors affecting a host nation's willingness and capacity to reliably secure, store, and provide access to prepositioned WRM assets. The factors that constitute SPRAT are summarized in Table 5.1. Although hardly exhaustive, this list of characteristics was designed to systematically provide planners with what we concluded would be the most pertinent political and strategic information about each of 195 potential host countries in developing prepositioning postures.

Table 5.1. SPRAT Measures

Measure	Considerations
Host-nation capacity and capability to secure and facilitate the movement of prepositioned WRM	
Internal stability of the host nation[a]	How stable are the host nation's governance structures as measured by levels of legitimacy, corruption, accountability, and rule of law?
Internal security of the host nation[a]	How vulnerable is the host nation to internal security threats (for example, insurgencies, coups, terrorism)?
External security of the host nation[a]	How vulnerable is the host nation to external security threats (for example, external conflicts, hostile neighbors)?
Host-nation infrastructure quality	How reliable is the quality of the host nation's ground, sea-based, railway, and air transportation infrastructure?

---

[9] Paul et al., 2013, pp. 32–35.

[10] Michael J. Lostumbo, Michael J. McNerney, Eric Peltz, Derek Eaton, David R. Frelinger, Victoria A. Greenfield, John Halliday, Patrick Mills, Bruce R. Nardulli, Stacie L. Pettyjohn, Jerry M. Sollinger, and Stephen M. Worman, *Overseas Basing of U.S. Military Forces: An Assessment of Relative Costs and Strategic Benefits*, Santa Monica, Calif.: RAND Corporation, RR-201-OSD, 2013; and Stacie L. Pettyjohn and Alan J. Vick, *The Posture Triangle: A New Framework for U.S. Air Force Global Presence*, Santa Monica, Calif.: RAND Corporation, RR-402-AF, 2013.

[11] Pettyjohn and Kavanagh, 2016.

Measure	Considerations
Host-nation climate conditions	How vulnerable is the host nation to major climate events and disruptions?
Host-nation willingness and reliability in storing and granting access to prepositioned WRM	
Existing WRM storage	Is WRM already stored within the host-nation territory?
Existing security agreement[a,b]	Does the United States have a mutual defense pact with the host nation?
Participation in major U.S.-led coalition operations	Did the host nation participate in major coalition contingency operations with U.S. forces since the terrorist attacks of September 11, 2001?
Host-nation political and economic cooperation with the United States[a]	How strong are political and economic ties between the United States and the host nation?
Host-nation security cooperation with Russia and China[a]	What percentage of arms shipments to the host nation come from Russia and China?
Host-nation economic cooperation with China	How strong are economic ties between the host nation and China, as measured by official and unofficial development as a percentage of gross domestic product (GDP), 2000–2014?
U.S. overseas presence	Does the United States maintain a major or permanent basing presence in host-nation territory?
Access challenges (peacetime)[c]	Has the host nation evicted, restricted, or otherwise challenged U.S. access to its territory during peacetime operations (1945–2014)?
Access challenges (contingency)[c]	Has the host nation denied, restricted, or otherwise challenged U.S. access to its territory during contingency operations (1945–2014)?

[a] Data in these measures are the same information (or updated raw data) as contained in the RAND Security Cooperation Prioritization and Propensity Matching Tool, November 2016 Excel spreadsheet. See Paul et al., 2013.
[b] Data are derived from Kavanagh, 2014.
[c] Data are from Pettyjohn and Kavanagh, 2016.

In one general category, planners are supplied with information about the capacity and capability of a candidate host nation to secure and facilitate the movement of prepositioned WRM for use in peacetime and contingency operations. Several of these measures, such as the internal stability of the host nation's regime, its vulnerability to internal and external security threats, and the reliability of its infrastructure, are consistent with other studies that use these factors as proxies for a host nation's suitability for foreign investment and asset prepositioning.[12] We also included a measure of a host nation's vulnerability to major climate events or environmental risks that could erode or destroy prepositioned equipment.[13]

---

[12] Measures for a government's internal stability and vulnerability to internal and external threats are derived from Jane's Military and Security Assessments Intelligence Centre, "Global Country Risk Ratings," IHS Markit, August 14, 2018. Assessments of a country's infrastructure were derived from World Bank, Logistics Performance Index, Washington, D.C., 2018. This index is an interactive benchmarking tool that rates the quality of a country's infrastructure based on a worldwide survey of logistics operators on the ground who provide feedback on the ease of operating in countries with which they trade.

[13] This measure was derived from Ursula Kaly, Craig Pratt, and Jonathan Mitchell, *The Demonstration Environmental Vulnerability Index (EVI) 2004*, Suva, Fiji: South Pacific Applied Geoscience Commission,

In a second general category, planners are supplied with information that we expected would shape a host country's willingness to store and grant access to prepositioned WRM assets. The first set of characteristics within this category are designed to capture elements of the nation's overall political and security relationship with the United States. The measures that serve as proxies for this relationship include whether the candidate host nation currently stores WRM assets on behalf of the United States;[14] whether it is a signatory to a mutual defense treaty with the United States;[15] and whether it meaningfully contributed in major recent U.S.-led coalition operations in Afghanistan, Iraq, and Libya and in the global coalition to counter the Islamic State.[16] The degree of political and economic cooperation between the United States and the candidate host nation is measured through an aggregation of several factors, including the volume of trade and foreign direct investment flows between the United States and the host nation; popular attitudes toward U.S. leadership in the host country; movement of travelers, immigrants, and exchange students between the two countries; and the degree to which a host nation's votes on United Nations General Assembly resolutions coincide with those of the United States.[17]

The second set of characteristics within this category aim to capture a candidate host nation's political, economic, and security relationships with Russia and China, the United States' near-peer competitors. If one assumes that the degree of political and economic cooperation with the United States informs a partner nation's propensity to cooperate in security affairs, then the same reasoning can be applied to assess a country's propensity to cooperate with the United States' strategic competitors. Understanding how near-peer competitors interact with—and potentially

---

Technical Report 384, 2004. The index is a multiyear effort to estimate each country's future risk of environmental shocks based on 50 climate and environmental indicators, such as sea temperatures, habitat loss, degradation rates, and frequency of natural disasters..

[14] Countries were identified as currently storing WRM assets according to a list of non-munitions WRM equipment data from the Air Force dated April 2018.

[15] Membership in mutual defense treaties was derived from a comprehensive database of U.S. security agreements and updated to reflect recent signatories and withdrawals from mutual defense alliances. See Kavanagh, 2014.

[16] Information regarding foreign partner contributions in support of the North Atlantic Treaty Organization (NATO)'s International Security Assistance Force and Resolute Support Mission in Afghanistan were derived from periodic summary reports of foreign troop deployments in Afghanistan archived at NATO, "Archive ISAF Placemats: NATO and Afghanistan," webpage, May 23, 2017a. Information regarding foreign troop contributions to Operation Iraqi Freedom were derived from Steven A. Carney, *Allied Participation in Operation Iraqi Freedom*, Washington, D.C.: Center for Military History, 2011. Information regarding country contributions to the 2011 intervention in Libya were derived from Ivo H. Daalder and James G. Stavridis, "NATO's Victory in Libya: The Right Way to Run an Intervention," *Foreign Affairs*, Vol. 91, No. 2, March/April 2012. Country participation in the 79-member Global Coalition Against Daesh was derived from the list of partners on the coalition's website, which was last accessed September 13, 2018 (Global Coalition Against Daesh, "Partners," webpage, undated).

[17] This composite measure is derived from an updated propensity score for "Category 3:US-PN Relationship" in the RAND Security Cooperation Prioritization and Propensity Matching Tool (Paul et al., 2013).. The measures that constitute the constructs in this category were reweighted and populated with updated data available as of August 2018. Data sources included the U.S. Bureau of Economic Analysis, the U.S. Census, Gallup, the U.S. Department of Homeland Security, the Institute for International Education, and the U.S. Department of State. See Appendix D.

exact leverage over—other countries is akin to the way that multinational firms seek to incorporate the behavior of competing suppliers in foreign markets.

In light of the dearth of trustworthy, publicly accessible data that can capture the full scope of the political, economic, and security relationship between host nations and the United States' near-peer competitors, SPRAT uses two reliable (if imperfect) data sources to approximate this relationship. First, SPRAT draws on global arms transfers data collected annually by the Stockholm International Peace Research Institute to measure the share of a host nation's arms imports that come from Russia and China.[18] Second, SPRAT evaluates what share of a host nation's GDP over the past decade is represented by official and unofficial Chinese development assistance. This construct draws from a data set, released by AidData, on official and unofficial Chinese development flows between 2000 and 2014.[19]

A third set of characteristics aims to provide planners with a summary of a candidate country's experience with hosting U.S. forces and any restrictions or denials of access it may have imposed on the United States historically during peacetime and contingency operations. Information on the current presence of U.S. forces overseas is derived from the declassified FY 2017 Base Structure Report, as well as the number of active duty military and permanent DoD civilian personnel deployments overseas, as reported by the Defense Manpower Data Center in June 2018.[20] SPRAT draws on the previously referenced research by Pettyjohn and Kavanagh, 2016, to report whether candidate host nations have ever imposed restrictions on or evicted U.S. forces stationed on their territory during peacetime and whether these countries have restricted or denied access to U.S. forces during contingency operations between 1945 and 2014.[21]

An Illustration of SPRAT's Utility

After we collected data on the 14 measures in Table 5.1 for each of the 195 candidate host nations, we developed a streamlined coding scheme that could present planners with a series of color-coded binary, ordinal, and categorical values to deliver a complex series of variables in a single, accessible, and visually simplified spreadsheet. Figure 5.1 displays a screenshot of a

---

[18] Stockholm International Peace Research Institute, "SIPRI Arms Transfers Database," webpage, undated.

[19] See Axel Dreher, Andreas Fuchs, Bradley Parks, Austin M. Strange, and Michael J. Tierney, *Aid, China, and Growth: Evidence from a New Global Development Finance Dataset*, Williamsburg, Va.: AidData, Working Paper 46, October 2017; and Austin M. Strange, Mengfan Cheng, Brooke Russell, Siddhartha Ghose, and Bradley Parks, *AidData Methodology: Tracking Underreported Financial Flows (TUFF)*, Version 1.3, Williamsburg, Va.: AidData, October 2017. These data have been used in research reports and peer-reviewed publications, including Austin M. Strange, Alex Dreher, Andreas Fuchs, Bradley Parks, and Michael Tierney, "Tracking Underreported Financial Flows: China's Development Assistance and the Aid-Conflict Nexus Revisited," *Journal of Conflict Resolution*, Vol. 61, No. 5, 2017.

[20] DoD, *Base Structure Report—Fiscal Year 2017 Baseline: A Summary of the Real Property Inventory*, Washington, D.C., April 4, 2018b; and Defense Manpower Data Center, "DoD Personnel, Workforce Reports & Publications," webpage, undated.

[21] Pettyjohn and Kavanagh, 2016, pp. 139, 152–158.

sample of countries and their scores on characteristics that represent a host nation's willingness to store and provide access to prepositioned WRM assets. (For a detailed explanation of the coding scheme and data sources for each measure, see Appendix D.) The 29 countries listed in Figure 5.1 represent the candidate overseas locations evaluated by PAF's 2010 study to derive a cost-optimized prepositioning posture of BEAR assets. Optimizing against total system costs generates a BEAR prepositioning posture that includes storage sites in nearly all of these countries.[22]

By supplementing this optimization analysis with systematic qualitative data about each potential host nation, SPRAT allows planners to more proficiently weigh the trade-offs between (1) greater cost and transportation efficiencies and (2) the strategic and political risks of prepositioning WRM assets within a host nation's borders. To illustrate these trade-offs, consider the strategic and political risks of prepositioning BEAR assets in two candidate countries, Djibouti and São Tomé and Príncipe. A strictly cost- and transportation-based optimization analysis would conclude that their respective geographic locations on the Horn of Africa and the western coast of Central Africa, respectively, make them compelling candidates to include in a global prepositioning posture.

From the standpoint of political and strategic risk considerations as presented by SPRAT, however, prepositioning WRM assets in these countries appears less desirable. As can be seen in the fourth column from the right in Figure 5.1, between 2000 and 2014, China has pledged or provided substantial development assistance to Djibouti and São Tomé and Príncipe in amounts that are equivalent to more than 3 percent of their overall GDP over the same period. And in both cases, these investments have translated into increased leverage that moves the countries closer to China's preferred political and security outcomes. In August 2017, China opened its first overseas military base in Djibouti.[23] And a year after a state-owned Chinese enterprise pledged to finance the development of an $800 million deep-sea port, São Tomé and Príncipe became the first in a wave of defections from diplomatic recognition of Taiwan.[24] Although cost-optimization may tempt planners to store WRM in these countries, the increasing leverage that China exacts over these nations may give planners pause as they weigh increased risks that national governments might deny access to prepositioned U.S. assets, and the use of local transportation might not be available as anticipated.

---

[22] McGarvey et al., 2010, pp. 23–25, 39–41.

[23] Ben Blanchard, "China Formally Opens First Overseas Military Base in Djibouti," Reuters, August 1, 2017.

[24] Karin Strohecker, "Sao Tome Signs Memorandum with China on Deep Sea Port," Reuters UK, October 12, 2015; and Louise Watt, "China Resumes Ties with Sao Tome in Triumph over Taiwan," Associated Press, December 26, 2016.

Figure 5.1. Excerpt from SPRAT

Country or Territory	ISO	FIPS	COCOM	EXISTING WRM STORAGE Is WRM already stored at HN site? (Yes/No)	EXISTING SECURITY AGREEMENT Does U.S. have a mutual defense pact with HN? (Yes/No)	PARTICIPATION IN MAJOR U.S.-LED COALITION OPERATIONS Did HN participate in major coalition contingency operations with U.S. since 9/11? (Yes/No)	WILLINGNESS/RELIABILITY U.S. POLITICAL/ ECONOMIC COOPERATION How strong are political and economic ties between U.S. and HN? (Low/Medium/High)	HN SECURITY COOPERATION WITH RUSSIA & CHINA What fraction of arms shipments to HN come from Russia and China? (None/Low/Medium/High)	HN ECONOMIC COOPERATION WITH CHINA How strong are economic ties between HN and China (as measured by official and unofficial development as % of GDP, 2000-2014)? (None/Low/Medium/High)	U.S. OVERSEAS PRESENCE Does U.S. maintain a major or permanent basing presence in HN territory? (Yes/No)
Afghanistan	0	AF	CENTCOM	No	No	Some	Medium	None	Low	Yes**
Azerbaijan	AZE	AJ	EUCOM	No	No	Some	Low	Low	Low	Yes
Bahrain	BHR	BA	CENTCOM	No	No	Some	Medium	None	None	Yes
Bulgaria	BGR	BU	EUCOM	No	Yes	All	High	None	Low	Yes
Cyprus	CYP	CY	EUCOM	No	No	Some	Medium	None	Low	Yes
Djibouti	DJI	DJ	AFRICOM	No	No	Some	Medium	Low	High	Yes
Ecuador	ECU	EC	SOUTHCOM	No	No	None	Medium	None	Medium	Yes
Greece	GRC	GR	EUCOM	No	Yes	Some	Medium	None	Low	Yes
Iraq	IRQ	IZ	CENTCOM	No	No	Some	Low	Low	Low	Yes**
Italy	ITA	IT	EUCOM	Yes	Yes	All	High	None	None	Yes
Japan	JPN	JA	PACOM	Yes	Yes	Some	High	None	None	Yes
Korea (Seoul). Republic of Korea	KOR	KS	PACOM	Yes	Yes	Some	High	None	None	Yes
Kuwait	KWT	KU	CENTCOM	No	No	Some	Medium	None	None	Yes
Luxembourg	LUX	LU	EUCOM	Yes	Yes	Some	High	None	None	Yes
Nigeria	NGA	NI	AFRICOM	No	No	Some	Medium	Medium	Low	Yes
Norway	NOR	NO	EUCOM	No	Yes	Some	High	None	None	Yes
Oman	OMN	MU	CENTCOM	Yes	No	Some	Medium	None	None	Yes
Panama	PAN	PM	SOUTHCOM	No	Yes	Some	Medium	None	None	Yes
Philippines	PHL	RP	PACOM	No	Yes	Some	Medium	None	Low	Yes
Qatar	QAT	QA	CENTCOM	Yes	No	Some	Low	Low	None	Yes*
Romania	ROM	RO	EUCOM	No	Yes	All	High	None	Low	Yes
Sao Tome and Principe	STP	TP	AFRICOM	No	No	None	Low	None	High	No
Senegal	SEN	SG	AFRICOM	No	No	None	Medium	Low	Low	Yes
Singapore	SGP	SN	PACOM	No	No	Some	Medium	None	None	Yes
South Africa	ZAF	SF	AFRICOM	No	No	None	Medium	Medium	Low	Yes
Spain	ESP	SP	EUCOM	No	Yes	All	High	None	None	Yes
Thailand	THA	TH	PACOM	No	No	Some	Medium	Low	Low	Yes
Turkey	TUR	TU	EUCOM	No	Yes	Some	Medium	None	Low	Yes
United Kingdom (Britain)	GBR	UK	EUCOM	Yes	Yes	All	High	None	None	Yes

NOTE: HN = host nation.

SPRAT's Limitations and Policy Implications

In its current form, SPRAT represents an initial prototype of a qualitative assessment tool that can help planners weigh the political and strategic risks of prepositioning WRM assets. Some may object that the broad categorical values of each of these measures are too oversimplified to accurately portray true underlying political and strategic risks. Others may desire a more systematic classification for measures that allow for country rankings across a single composite risk metric. Moreover, SPRAT reflects the shortcomings of the data it seeks to supply. Gathering open-source data that are systematically available for 195 countries is a considerable feat; measures, then, may be imperfect proxies for the characteristics they seek to represent.

However, from the perspective of A4 planners, who often have to make decisions in a short time frame with limited access to information, SPRAT can, at the very least, provide a general overview of political and strategic risks that planners might not otherwise have for WRM prepositioning analyses. We recommend continuing to expand and validate the data that constitute SPRAT's framework in future iterations of the tool, including the addition of information not available to the general public, as well as coordinating with the Office of the Under Secretary of the Air Force for International Affairs and the U.S. State Department for information they may have available. Even without those enhancements, SPRAT could help the Air Force incorporate political and strategic considerations more systematically, particularly in operationally limited or denied environments in which the United States confronts near-peer competitors.

Chapter Six
Conclusions and Recommendations

Air Force WRM is not truly globally managed today. Over the past decade, the Air Force has taken steps to centralize WRM support by establishing the 635 SCOW. The SCOW maintains centralized acquisition, distribution, and inventory for a small number of Air Force assets, BEAR, fuel support equipment, and other support equipment. Even for those assets, the SCOW does not control positioning or maintenance, and it has no oversight of WRM beyond those classes of supply. The current system is positioned more for efficiency than effectiveness. Establishing the 635 SCOW with WRM responsibility and authority is a step toward centralization; however, there are process, tool, and system enhancements that could further centralize the management of WRM (with decentralized execution) to better support strategic resiliency and responsiveness goals.

The CDO environment expected in the future will likely create new risks and uncertainties, and such an environment is better served through a centralized management approach. When policies are enforced and adhered to, centralized management processes yield standardization. An Air Force global manager of WRM could globally manage capabilities, moving away from focusing on national stock numbers and item nomenclature (for example, generators, tents, and trucks) and toward discussing WRM capabilities (such as WRM needed to support the ability to receive forces at a base, project force, and recover the base) with a view toward effectiveness. Establishing DABS kits is a step toward standardization and centralization. Such systems include nontraditional WRM and provide a capability. Yet DABS kits are not standardized. Standardizing a DABS and making it more tailorable would allow each AOR to take a standardized capability and gear that capability toward the AOR's specific requirements.

Next, we offer some recommendations for methods that a global manager of WRM could employ to improve centralized or global WRM management. First, **a global manager of WRM should standardize and validate warfighter demands**. The Air Force has some systems used today to validate some but not all functional requirements (for example, BEAR, POL, and munitions). In this analysis, we demonstrated the use of Lean-START to generate and compare requirements across AORs, demonstrating the type and utility of a tool needed by a global manager of WRM to validate warfighter requirements beyond the few individual systems the Air Force currently has.

As discussed earlier, **the Air Force should move away from inventory management and toward capability management for mission support assets**. In this analysis, we created buckets of WRM resources and capabilities aligned with a base's different activities over different time periods. These buckets provide a way for a global manager to think about which capabilities (and, at a lower level of detail, which resources) need to be prepositioned.

Once a global manager knows which capabilities need to be prepositioned, he or she should prioritize those capabilities. Thus, **the Air Force should develop a method to determine priority for WRM prepositioning.** In this analysis, we developed a decision tree that evaluates the temporal order of each WRM capability and aligns that order (as well as other aspects, such as transportability and cost) to a prepositioning posture. The decision tree we developed is an example of the type of repeatable process that a global manager could use to think about how to posture the most time-critical WRM capabilities and which posture could be used for less-critical capabilities.

The Air Force should optimize its WRM prepositioning posture. PAF researchers developed a tool, the PRePO model, to answer strategic-level questions about prepositioning WRM. The model was designed to find optimal solutions—to, for example, lowering costs, minimizing the time it takes for a base to receive its required WRM, or evaluating user-developed postures—based on storage allocations among candidate sites, costs of storage and transportation, and a transportation schedule. In this analysis, we used the PRePO model to demonstrate the types of analyses a global manager of WRM should be conducting. As an example, we assessed how long it would take for a set of WRM storage COAs to make a beddown of operating locations initially capable for operations for conflicts in four different global regions. Using this example, we were able to demonstrate how operational effectiveness could be improved using different WRM prepositioning strategies. Decision support tools like the PRePO model can show decisionmakers the costs and risks associated with various positioning strategies.

To supplement the PRePO analysis, we developed SPRAT, a tool that considers political, economic, and strategic factors beyond what is modeled in the PRePO tool. Such qualitative factors as a host nation's long-term relationship with the United States and its internal political and economic stability can be informative when thinking about where to preposition WRM assets. **The Air Force should consider political factors when analyzing WRM prepositioning strategies**. A planner can use the political considerations to influence the inputs into the PRePO model.

Finally, we recommend that the Air Force **adopt tools and metrics to measure readiness for mission support assets**. As found by the Air Force's Sustainment Review Team in 2018, reporting on WRM equipment readiness is inconsistent across commands and may not always accurately reflect equipment condition. The Air Force reports aircraft and supply support readiness in the SORTS. But the Air Force needs a way to regularly report its ability to provide mission support capabilities, not item inventories—perhaps in SORTS or DRRS.

To achieve true global management of WRM, with the capabilities recommended here, the Air Force will have to invest in an organization with adequate resourcing (manpower, facilities, tools, and system), authorities, and governance to buy WRM capabilities, sustain them, and make trade-off decisions about their global positioning. This would be a cultural shift from how WRM is managed today.

Global Management Case Studies

In this appendix, we present five case studies of global management practices outside the U.S. Air Force. First, from industry, we investigate the management of oil and gas drilling. Next, we examine FEMA as an example of a government agency employing global management strategies. Finally, we evaluate other military services' WRM management practices. We use these case studies to help determine sound management practices that can be used in differing situations.

Industry Example: Oil and Gas Drilling

Oil and gas drilling is the process of boring holes into the earth to extract oil and gas for production. Companies (for example, Helmerich and Payne, Transocean, Patterson UTI, and Diamond Offshore) engage in this activity typically on a contract basis on the behalf of oil exploration and production corporations (such as, ExxonMobil, Shell, or BP).[1] A client oil exploration and production company will hire a drilling company to mobilize equipment to the drilling site to engage in operations.[2] Contracts specify where and when the drilling company must drill and the desired depth of the well.[3] Ground vehicles, freighter ships, helicopters, and cargo aircraft are used to transport assets across sites. Drilling companies will also rely on and source materials from many vendors (for example, more than 400 vendors for offshore platforms). Oil and gas drilling may use various material and transportation assets, depending on the environments where drilling equipment will be deployed.

Drilling Operations

Although oil and gas drilling companies are for-profit corporations, some of their activities and complications do bear resemblance to Air Force WRM management.[4] The key systems

---

[1] We reviewed the corporate websites and 10-K (financial performance) reports for the largest oil and gas drilling companies in the United States.

[2] Common equipment includes drilling rigs, jack-up rigs, drill-ships (offshore), semi-submersible platforms (offshore), and tubular goods (pipes, tubing, and casing).

[3] Oil and gas drilling services may include well drilling, which involves a drill boring vertically into the ground; directional drilling, which is the practice of drilling nonvertical wells; and reconditioning, which involves the reactivation of a past drill site. Individual companies will typically specialize in either offshore (ocean) or land (onshore) drilling. See "Markets: Oil and Gas Drilling," *New York Times*, undated.

[4] Christopher M. Chima, "Supply-Chain Management Issues in the Oil and Gas Industry," *Journal of Business & Economics Research*, Vol. 5, No. 6, June 2007. Claus Steinle and Holger Schiele, "Limits to Global Sourcing? Strategic Consequences of Dependency on International Suppliers: Cluster Theory, Resource-Based View and Case Studies," *Journal of Purchasing and Supply Management*, Vol. 14, No. 1, March 2008; Gurdeep Singh, Anant

involved are the drilling rigs. They are deployed to forward locations, where they need to be ready to perform in austere environments and where the oil and gas company hopes to put them to use, because idle rigs could result in losses of thousands of dollars each day.[5]

Domestic land drilling best illustrates the similarities between asset management in this industry and the Air Force's WRM management. Figure A.1 shows the *oil plays* (that is, the groups of oil fields or prospective oil fields in the same region) and land drill locations in CONUS in February 2014. Oil plays are analogous to AORs in the WRM context. Furthermore, we can view the drill sites as being analogous to FOLs. A drilling company first transports a drill to a site within the oil play, then sets up the rig, which then drills to the desired depth.[6] All equipment necessary to operate the drilling rig comes with it. Usually, the drilling rig is then replaced by another (service) rig that is tasked with pumping and extracting the oil, and the drilling rig is taken down and transported to another site.[7] Offshore drilling works on similar principles, in which drill ships and platforms are moved across different sites.

Figure A.1. Oil Plays and Rig Locations in the United States, February 2014

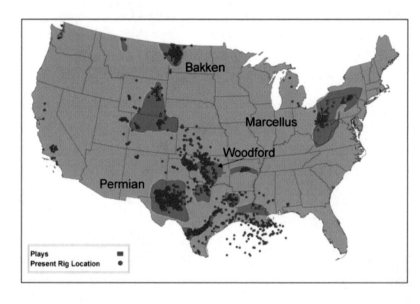

SOURCE: David Wogan, "Oil and Natural Gas Drilling Rigs Are Moving In at a Furious Pace," *Scientific American*, March 5, 2014.

Tripathi, Anupriya Srivastava, and Mahesh Iyer, "Integrated Supply Chain Outsourcing: Expanding the Role of Third Party Logistics in the Upstream Industry," Society of Petroleum Engineers Oil & Gas India Conference and Exhibition, November 24–26, 2015; and Patrick Scott and Boxi Xu, *Multi-Echelon Inventory Modeling and Supply Redesign*, Cambridge, Mass.: Massachusetts Institute of Technology, 2017.

[5] John Kemp, "U.S. Drilling Costs Start to Rise as Rig Count Climbs: Kemp," Reuters, July 14, 2017.

[6] Rick Carr and Sam Pearson, "Unconventional Drilling Requires Managing Transportation Logistics," *Oil and Gas Journal*, June 4, 2012.

[7] This process (transport, set up, drill, take down) takes two to four weeks for each drill.

Management and Organizational Practices

Operating land drills and deploying related equipment can perhaps best be compared to employing a distributed WRM posture. There is a less formal distinction between sustainment activities and materiel management in the oil and gas industry. Companies use yards and facilities within each oil play (onshore) or near ports (offshore) to store additional material not at the rigs, including repairables and spare parts. These regional facilities provide for sustainment of drilling equipment and may be considered as support locations; however, besides keeping specialized equipment for the drill, companies engaged in onshore drilling will either source other requirements from corporate vendors or have on-site management locally source items to satisfy requirements. Other requirements for offshore operators are typically handled by global or regional arrangement with vendors; barges and helicopters are used to transport items to offshore rigs, and some companies use containerized cargo to reduce costs.

The organizational structures of these companies reveal that, after a division of management based on onshore versus offshore drilling sites, subordinated control of drilling operations is mainly structured by region. Movement of individual rigs and equipment within a region is directed by regional and local management, and any inter-regional realignment would be handled by the regional managers involved. Regional managers are senior executives in these organizations.

Demand Management

Oil and gas drilling contracts are normally region-focused; state how much the drilling company will be paid; and include negotiated terms on the necessary lead times, costs, delivery dates, and so forth. It is a highly competitive, demand-driven industry.

Demand for drilling services in each region is sporadic, with some localized areas that often experience a flurry of activity. High energy prices can trigger simultaneous demand across multiple regions; as a result, supply chains can become increasingly strained, and lead times may rise. Information technology services enable the recognition of demand requirements and subsequent coordination. For example, some companies use proprietary software to manage rig realignments, and on-site locations use commercial software (for example, SAP) to automatically reorder additional sustainment material from facilities.

Regional managers typically control rig realignment and facilities in their region and, as high-level executives, meet with each other regularly (for example, weekly) and with senior corporate officers (for example, a vice president of operations and a chief executive officer) on occasion. These managers can direct regional staff to coordinate logistics and sourcing and strategize with supply chain executives.

Similarities to WRM Management

If drill sites and regional oil plays are to be interpreted as forward locations and AORs, respectively, then the oil and gas drilling industry adopts something akin to a distributed posture. For instance, essential equipment travels with the drilling rig. And the industry avoids transporting equipment outside the region and tries not to stockpile equipment, so there is little excess capacity. Depending on the organization, stocks of repairables and spare parts are held at centralized or regional locations, with most other requirements being met from established vendors or procured sources. It is unknown what is the best practice for this industry, and companies continue to revise their organizational structures and update their practices.[8]

The demand signal is unpredictable and dependent on energy prices. Because the sites in an oil play that are the cheapest to exploit change over time and because the cost to extract varies across plays, the demand signal is usually region-specific. However, when high energy prices result in simultaneous activity in all regions, analogously to simultaneous operations across AORs, it places additional strain on both sustainment and transportation of the drilling systems.

Government Example: Federal Emergency Management Agency

We reviewed the evolution of FEMA from various academic studies and the FEMA website to understand how a nonmilitary government agency is structured, prepositions goods, and delivers those goods in a complex system. FEMA's stated objective is to posture its organization and "the whole community to provide life-saving and life-sustaining commodities, equipment, and personnel from all available sources."[9] A critical function of this goal is the strategic placement and delivery of a range of emergency items, such as food, water, ice, blankets, medicine, evacuation equipment, and temporary housing modules. Timely delivery requires effective cooperation with layers of municipal, county, and state-level government agencies.

FEMA faced scrutiny for its response to and management of the effects of Hurricane Katrina in 2005. Critics cited poor preposition planning and failed delivery of vital, life-saving goods as contributing to the overall failure of the agency to respond. FEMA identified the lack of an established and tested communication channel from various localities (overlapping municipal, fire, and police districts) to the federal level as a primary factor that prevented prepositioned items from making it to those in need.

Congressional legislation in 2007 mandated that FEMA conduct a strategic reorganization of its processes to address some of these issues. The new organizational structure (see Figure A.2) is a hub-and-spoke distribution and evacuation system divided into ten regions. Each regional office acts as the hub to connect the various spokes (layers of local and state government) to the many federal-level agencies that constitute the whole domestic security and human welfare

---

[8] Chima, 2007.

[9] FEMA, *2018–2022 Strategic Plan*, Washington, D.C., March 15, 2018, p. 20.

provision system. Although studies still suggest areas for improvement, the strategic communication and delivery structure has improved preposition and delivery.

Figure A.2. FEMA Alignment into Distribution and Evacuation Regions

SOURCE: U.S. Department of Homeland Security, *National Response Framework*, 2nd ed., Washington, D.C., May 2013, Figure 5.

FEMA is only one among many national-level agencies and sub-offices that contribute to emergency management (the Centers for Disease Control and Prevention is another). FEMA's 2018–2022 Strategic Plan emphasizes the need to improve on strategic management and coordination of supplies with the many local, state, and federal stakeholders involved in the process:

> No one agency or organization is able to store enough equipment, supplies, and commodities, or marshal enough internal teams to quickly and fully meet all the needs . . . . This necessitates the collective assets and contributions of many supporting organizations aligned through a coordinated structure . . . . Consequently, FEMA must consider the appropriate balance of what it keeps on hand in warehouses, what it can quickly acquire through contracts, and what can be provided by volunteer organizations, other Federal agencies, [state, local, tribal, and territorial governments], and the private sector.[10]

---

[10] Federal Emergency Management Agency, 2018, p. 24.

Similarities to WRM Management

FEMA's distribution and evacuation regions are similar to military AORs, and, like in the military, the demand for FEMA resources is unpredictable and usually regionally focused. In its ten regions, FEMA maintains what could be considered a distributed posture to work closely with state, local, and municipal levels of government. FEMA also maintains a more centralized organizational structure at higher levels of the organization to provide coordination across the regions, which is similar to the role of a WRM global manager.

Other Services' WRM Management

Army

Army prepositioned stocks (APS) are forward-positioned stocks of materiel that are designed to reduce the transportation burden required to support forward deployment of Army personnel and to sustain units in the initial days of a conflict until supply lines are established.[11] APS are located worldwide and, in some cases, afloat.[12] Table A.1 outlines the global footprint of APS.

Table A.1. APS Locations

Name	Location
APS-1	CONUS
APS-2	Europe and Africa
APS-3	Afloat
APS-4	Northeast Asia and Pacific
APS-5	Southwest Asia
APS-6	Central America/South America/Caribbean

SOURCE: HQDA, 2015.

There are five types of APS: unit sets, Army war reserve sustainment stocks, operational project stocks, war reserve stocks for allies, and activity sets. Unit sets and Army war reserve sustainment stocks are the initial materiel, equipment, major end items, and secondary items that units may fall onto following deployment, and that are used to provide the minimum required sustainment until supply lines are established. Sustainment stocks may also be used for resupply following combat loss. Operational project stocks include materiel above a unit's table of organization and equipment, or the equipment included in unit sets, to support potentially multiple Army plans or contingencies. War reserve stocks for allies is an Office of the Secretary of Defense program to provide assistance to allies during war. Finally, activity sets are intended

---

[11] Jacqueline Georlett and Bruce Daasch, "Army Pre-Positioned Stocks Support Army Readiness," *Army Sustainment*, May–June 2017.

[12] Headquarters, Department of the Army (HQDA), *Army Pre-Positioned Operations*, Army Techniques Publication 3-35.1, Washington, D.C., October 2015.

for use during deployment for training exercises outside CONUS and typically include a smaller subset of the materiel included in unit sets.[13]

APS Management Authorities

All APS are owned by HQDA, which develops funding requirements and serves as the release authority for APS.[14] Management of materiel varies by materiel type, and medical materiel (Class VIII) is distinct from all other materiel types, as summarized in Table A.2. The managing authority is responsible for storage, maintenance, and issue of APS, among other things.

Table A.2. APS Management and Authorities

Responsibility	Class VIII Materiel	All Other Materiel
Owner	HQDA	HQDA
Release authority	Secretary of Defense; HQDA (G 3/5/7)	Secretary of Defense; HQDA (G-3/5/7)
Funding requirements	The Office of the Surgeon General is the executive agent for HQDA	The U.S. Army Materiel Command is the executive agent for HQDA
Management	The United States Army Materiel Management Agency is the executive agent for the Office of the Surgeon General	The Army Sustainment Command is the executive agent for the U.S. Army Materiel Command

SOURCE: HQDA, 2015, Chapter 2.

Global Management of APS in Doctrine but Not in Practice

According to Army doctrine, APS are not aligned to any specific contingency, unit, or geographic area. Army Techniques Publication 3-35.1 states, "APS is not dedicated to specific units or theaters, but can be issued to units whenever and wherever required as directed by the Secretary of Defense,"[15] and "APS are national assets, owned by the Department of the Army, and, when issued, stock-funded items stored in APS are required to be purchased by the receiving [Army forces]."[16] Indeed, during Operation Iraqi Freedom, equipment for deploying troops was drawn from multiple APS sites, including the regionally aligned site (APS-5 and APS-3).[17]

Although Army doctrine outlines a global management strategy for APS, logistics limitations make the likelihood of global management in practice—where materiel in one geographic APS

---

[13] HQDA, 2015.

[14] HQDA, 2015. HQDA coordinates with the Secretary of Defense for release of APS during major contingencies, but it has the authority to independently release APS materiel for small-scale operations or national emergencies.

[15] HQDA, 2015, para. 1-6.

[16] HQDA, 2015, para. 4-1.

[17] HQDA, 2015; and Michael J. Harlan, *Force Projection Logistics Atrophy: Affliction and Treatment*, Carlisle Barracks, Pa.: U.S. Army War College, March 2011.

site is moved to a more distant geographic site—very low. During Operation Iraqi Freedom, when materiel from multiple APS sites was used, there were numerous logistical issues that caused delays in receipt of materiel; such issues were particularly related to transportation and timelines for release of materiel from APS.[18] More-recent studies have similarly highlighted the significant challenges in releasing materiel and moving that materiel, even within the theater where that materiel was stored, from APS sites in a timely manner.[19]

Reflecting the reality of the logistical challenge of transporting often heavy Army equipment, recent efforts have begun to operationalize materiel stored in APS sites.[20] In this effort, the unit sets and Army war reserve sustainment stocks that are stored at a geographic location are modified to be more suited for contingencies that may arise in that geographic area. Although this is likely to speed the release of materiel from APS sites and will ensure that the materiel that troops fall into requires little to no modification for a contingency in that geographic location, it is now even more unlikely that materiel stored in one geographic area will be drawn upon to fill a need in another.

Navy

Until 2011, the Navy maintained a WRM program and specific instructions for WRM organization and management.[21] The purpose of this program was to "to provide the additional materiel, over and above peacetime operating and training stocks, needed to support the force structure dictated by the [Secretary of Defense] planning guidance." Responsibility for WRM management was subordinated to a resource sponsor in the OPNAV and a war reserve project manager. The resource sponsor was responsible for WRM project approval and funding, and the project manager carried the responsibility for both "designing the project in response to component commander needs and maintaining the allowance configurations." Overall guidance and direction for the WRM programs was provided by the Deputy Chief of Naval Operations for Logistics, Supply Programs and Policy Division. Sourcing for WRM came from "peacetime operating stocks, training stocks, commercial contracts, host-nation support agreements, and bilateral military agreements." If these sources were insufficient to supply needs, "the

---

[18] Eric Peltz, Marc L. Robbins, Kenneth J. Girardini, Rick Eden, John M. Halliday, and Jeffrey Angers, *Sustainment of Army Forces in Operation Iraqi Freedom: Major Findings and Recommendations*, Santa Monica, Calif.: RAND Corporation, MG-342-A, 2005; Harlan, 2011.

[19] Elvira N. Loredo, Ryan Schwankhart, Abby Doll, Karlyn D. Stanley, Marta Kepe, Bryan Boling, Luke Muggy, Endy M. Daehner, and Simon Véronneau, *Challenges to Timely Movement of Forces in Europe: Moving Heavy Equipment, Ammunition, and Fuel*, Santa Monica, Calif.: RAND Corporation, 2019, Not available to the general public; and Elvira N. Loredo, Ryan Schwankart, Michelle D. Ziegler, Luke Muggy, Abby Doll, John Halliday, and Karlyn D. Stanley, unpublished RAND Corporation research, 2018.

[20] E. Daly and T. Fore, *Operationalizing Army Pre-Positioned Stocks (APS)*, Rock Island Arsenal, Ill.: U.S. Army Sustainment Command, March 30, 2017, Not available to the general public.

[21] Office of the Chief of Naval Operations (OPNAV), *Navy War Reserve Materiel Program*, Office of the Chief of Naval Operations Instruction 4080.11D, January 21, 1999.

Navy/Marine Corps component commanders [could] request that they be addressed by establishment of a WRM Project." WRM stocks were usually "held in the supply system as 'swing stocks,'"—that is, stocks that could be used in any scenario. This method of positioning WRM stocks allows for a reduction of costs (procurement, storage, handline, and maintenance) and facilitates replacement of aging stock. However, some WRM projects, with materiel urgently needed in possible operations, could be positioned in theater as "starter stock."

The WRM program instruction was cancelled by the Deputy Chief of Naval Operations for Logistics action memo dated October 26, 2011. The reason for cancellation was given as follows: "NO LONGER RELEVANT Instruction as written is neither applicable nor executable as a reference."[22] Furthermore, it was noted that the Navy no longer maintained WRM and that Naval Supply Systems Command deleted WRM project management codes effective FY 2006.[23] Additionally, the WRM requirements guidance, Naval Supply Systems Command Instruction 717, was canceled in 2012 with the note, "Function transferred and War Reserve Rqmts Deleted."[24] Some of these functions may have been transferred to the Defense Logistics Agency. As of September 2018, this instruction has not been replaced. The Navy Military Sealift Command controls and maintains prepositioning ships that carry materiel for the Air Force, Army, and Marine Corps, but it does not decide on which cargo or when or where the cargo is delivered. According to our review of the Department of the Navy's spares and ordnance procurement report for FY 2018, no budget line items for war reserve or prepositioned materiel are apparent, indicating that no programs exist.[25] Thus, the Navy has mostly, if not entirely, removed itself from WRM-oriented programs.

Marine Corps

Marine Corps WRM is managed by Blount Island Command at the Marine Corps Support Facility Blount Island in Jacksonville, Florida. The command oversees the storage, maintenance, and updating of WRM assets stored either in Norway or on one of 12 afloat prepositioning ships. The equipment and supplies stored in Norway and maintained by the Norwegian government as part of the country's NATO contribution are intended for use in the U.S. European Command AOR. The 12 afloat prepositioning ships are intended to support Marine Corps expeditionary

---

[22] Directives, Forms and Report Management Office, Director Navy Staff, Office of the Chief of Naval Operations, email correspondence with the authors, April 2018.

[23] It was also stated that OPNAV will revise the OPNAV Instruction 4040.39 series *Navy Expeditionary Table of Allowances and Advanced Base Functional Component Policy* to incorporate applicable elements from OPNAV Instruction 4080.11D, including Navy Expeditionary Tables of Allowance, Advanced Base Functional Components, and Navy prepositioned material (formerly Navy WRM). The Logistics, Supply Programs, and Policy Division will remain engaged with OPNAV Instruction 4040.39 series development and coordination.

[24] Office of Corporate Communication, Director Navy Staff, Office of the Chief of Naval Operations, Naval Supply Systems Command, email correspondence with the authors, April 2018.

[25] Department of the Navy, "Budget Materials," webpage, undated.

operations worldwide. Blount Island Command has a goal of trying to get each ship into Blount Island every 37–43 months, where the ships are off-loaded and all equipment stored on board is inspected, repaired, and refurbished. Once the equipment is off-loaded, the empty ship then sails to a commercial shipyard, where the hull is inspected and other refurbishment takes place. Approximately 90 days after the equipment is initially off-loaded, the ship returns to Blount Island, the equipment and supplies are reloaded in a few days, and the ship returns to sea. Although many of the ships share some similar equipment, some have very specific capabilities—for example, lighterage that allows ships to be off-loaded without being in a port, as well as hose assemblies that allow tanker ships to off-load fuel without going into port. The Marine Corps trains for the off-loading of ships and use of WRM regularly during theater-level exercises and during humanitarian assistance and disaster relief operations.

Conclusions

Most organizations choose a management strategy based on the organization's goals, anticipated risk, and uncertainty. As noted in Chapter Two, whereas industry's top priority is usually profit, the Air Force has a different motivation: defending the United States and its interests. As a result, the Air Force is often required to balance the ensured effectiveness of an operation (often associated with decentralized management practices) with efficiency in the systems used to support that operation (which typically leads to centralized management practices); thus, there likely is no one best management practice for the Air Force.

Appendix B
A Visualization Tool to Map WRM Processes

The WRM enterprise is large and complex. In this appendix, we explain how we developed a database and visualization tool to illustrate the relationships and the degree of connectedness between various WRM processes and organizations.

Initially, we relied on AFIs—mainly AFI 23-101 (*Air Force Material Management*) and AFI 25-101 (*Air Force War Reserve Materiel (WRM) Policies and Guidance*)—and discussions with Air Force WRM SMEs to develop a high-level view of the inherently complex process.[1] Although the discussions with SMEs provided some understanding of key offices and functions, they did not provide the desired degree of dependency, or connectedness, between the offices. The goal of this analysis was to understand the WRM system as a whole. Using the many roles and processes outlined in AFI 23-101 and AFI 25-101, we organized the data to conduct a network analysis. See Figure B.1 for an example of the data used from AFI 25-101.

Figure B.1. Excerpt from AFI 25-101, Chapter 2, "Roles and Responsibilities"

Chapter 2

ROLES AND RESPONSIBILITIES

2.1. Deputy Chief of Staff, Logistics, Installations, and Mission Support: (AF/A4).

2.1.1. Assists the Secretary of the Air Force (SECAF), other Secretariat offices, and the Chief of Staff in carrying out the training, organizing, and equipping of personnel for all facets of Logistics, Installations, and Mission Support programs of the Department of the Air Force to include the WRM program.

2.1.2. Exercises authority relating to WRM requirements and positioning, delegated to the SECAF pursuant to DoDI 3110.06, *War Reserve Materiel (WRM) Policy*.

2.2. Director of Logistics (AF/A4L).

2.2.1. Develops WRM policy and guidance to include AFPD 25-1, *War Reserve Materiel*, and this instruction, while ensuring WRM guidance in functional area instructions does not conflict.

2.2.2. Develops long-term enterprise level logistics operations strategy, concept of operations, tactics, techniques, and procedures.

2.2.3. Implements logistics policy for petroleum products, equipment, aircraft maintenance and munitions.

2.2.4. Ensures WRM objectives are consistent with Office of Secretary of Defense and Joint Chiefs of Staff (JCS) strategic guidance.

2.3. Logistics Operations, Plans and Programs Division (AF/A4LX).

SOURCE: Department of the Air Force, 2015, Chapter 2.

---

[1] Department of the Air Force, 2015; and Department of the Air Force, *Air Force Materiel Management*, Air Force Instruction 23-101, December 12, 2016.

To develop the network analysis database, we translated the information from both AFIs, by chapter, listing role or organization by row, with the associated descriptions—process, staff and organizations involved, function, and interactions and cross-communication with other offices—horizontally across fields or cells. Figure B.2 shows an example of the database using information from AFI 25-101.

Figure B.2. Excerpt from the RAND-Developed Database of WRM Roles and Processes

Code	Title	Organization	AFI-25-101 Descriptor 3	AFI-25-101 Descriptor 4	AFI-25-101 Descriptor 5	AFI-25-101 Descriptor 6	AFI-25-101 Descriptor 7	AFI-25-101 Descriptor 8	AFI-25-101 Descriptor 9
2.1.	Deputy Chief of Staff, Logistic	AF/A4	Assists the Secretary of the Air Force (SECAF), other Secretariat offices, and the Chief of Staff in carrying out the training, organizing, and equipping of personnel for all	Exercises authority relating to WRM requirements and positioning, delegated to the SECAF pursuant to DoDI 3110.06, War Reserve Materiel (WRM) Policy.					
2.2.	Director of Logistics	AF/A4L	Develops WRM policy and guidance to include AFPD 25-1, War Reserve Materiel, and this instruction, while ensuring WRM guidance in functional area instructions	Develops long-term enterprise level logistics operations strategy, concept of operations, tactics, techniques, and procedures.	Implements logistics policy for petroleum products, equipment, aircraft maintenance and munitions.	Ensures WRM objectives are consistent with Office of Secretary of Defense and Joint Chiefs of Staff (JCS) strategic guidance.			
2.3	Logistics Operations, Plans and Programs Division	AF/A4LX	Develops and publishes AFPD 25-1 and AFI 25-101. Ensures applicable WRM guidance is consistent with other functional area instructions.	Provides inputs to the War and Mobilization Plan (WMP) as required.	Ensures WRM is incorporated in the base support planning process as defined by AFI 10-404, Air Force Operations Planning and Execution.	Manages the Logistics Module (LOGMOD) system which provides the web-based capability for deployment and reception planning, execution to	Responsible for maintaining Base Support and Expeditionary global visibility of resources, to include WRM, at potential forward operating locations for		
2.4.	Logistics Readiness Division	AF/A4LR	Acts as the Functional Area Manager (FAM) for vehicles, petroleum products, and fuels related equipment to include FSE.	Provides policy guidance, direction and oversees policy implementation for managed WRM.	Oversees and manages Air Force Equipment Management System (AFEMS) IAW AFI 23-101, Air Force Materiel Management, and this instruction.	Develops policy and guidance for effective operation and official use of Government Motor Vehicles to include WRM vehicles IAW AFI 24-301, Vehicle Operations, AFI	Coordinates with the Defense Logistics Agency Energy (DLA Energy) concerning management, acquisition, transportation, storage, inventory accounting,	Advocates for WRM stock fund programming and allocation.	Develops policy and guidance for materiel management processes associated with WRM.
2.5.	Air Force Element Vehicle and	AFELM VEMSO	Manages WRM vehicle assets (as MAJCOM functional management) IAW AFI 24- 302, Vehicle Management, and this publication.	Assigns primary WRM vehicle management functions within the headquarters. AFELM VEMSO will work with the 635 SCOW/WM	Manages enterprise-level fleet support information technological programs, e.g., Logistics Information Management System-Enterprise View (LIMS-EV), e.g., AFEMS and Standard	Acts as the focal point for WRM fleet managers in the execution of vehicle management transactions across multiple AF systems,	Provides liaison between installation WRM managers and AFLCMC/WNZ Support Equipment & Vehicles Division concerning registered vehicle/vehicular	Serves as the enterprise authority for all vehicle procurements. Appropriations will be distributed to AFELM VEMSO for procurement	
2.6.	Air Force Petroleum	AFPET	Manages the AF fuel and equipment programs to include the AS.	Manages FORCE acquisition in coordination with the WRM GM.	Validates, consolidates and coordinates annual fuel requirements with DLA Energy.	As appropriate, notifies the affiliated Combatant Command/Joint Petroleum Office, Sub-Area Petroleum Offices, DLA Energy Regional/Field Office, when			

After the information was entered into the database, we pulled out keywords, offices, and tools, as outlined in the descriptions. We coded a system that would identify and tag each of these key items per their relative descriptions. This output was entered into Gephi, a visualization software, which produced a first iteration that was an incomprehensible network map. With input from SMEs, we identified and attributed weights to the most relevant offices, tools, and keywords in the WRM process (1 = least important, 5 = most important) in the AFIs. We separated the terms into the following categories:

- organizations, offices, and working groups
- systems, tools, and guidance
- verbs and keywords.

Each of these terms is represented by a *node* (a circle) that is connected to other nodes by an *edge* (a line or lines). The size and intensity of the color or the darkness of the node denotes the degree of interconnectedness, or how many relationships that node has with other nodes. The size or thickness of the edge denotes the weight or depth of the relationship between two nodes. See Table B.1 for an explanation of the visualization diagrams.

Table B.1. Explanation of the Visualization Diagrams

Node Label	Name of the Organization, Tool, or Verb That Appears in the AFI and Has an Established Relationship
Node size and intensity (darkness of color)	Interconnectedness of node—how many relationships it has with other nodes The larger and darker the node, the greater the degree of interconnectedness
Edge thickness	The weight attributed to the relationship between any two nodes
Edge label	The code from the AFI in which the relationship between organizations, tools, and verbs is established
Color group (yellow, brown, red)	Different communities—or organizations and offices that interact with one another with more frequency than others—are denoted by distinct color groups

We attributed weights to the relationships to "connect the dots" with respect to five distinct combinations of the three categories:

- Set 1: Organizations/offices/working groups to organizations/offices/working groups
- Set 2: Organizations/offices/working groups to systems/tools/guidance
- Set 3: Organizations/offices/working groups to organizations/offices/working groups with systems/tools/guidance (as edges, or connecting lines)
- Set 4: Organizations/offices/working groups to verbs/keywords
- Set 5: Organizations/offices/working groups to organizations/offices/working groups with verbs/keywords (as edges, or connecting lines).[2]

For each of these sets, we created a matrix of unique pairwise edge-lists. The matrix contains the weight of each edge and a label that identifies the AFI and section number in which the relationship is established. Each connection between and among these items is identified by a unique identification (ID) and its associated label where the connection is made in the text of the AFI. See Figure B.3 for an excerpt from an edge-list matrix.

Figure B.3. Sample Edge-List Matrix Illustrating How the Label Links the AFI to the Connection Between Nodes

Source	Target	label	type	id	timeset	weight
49_MMG	AFCEC	{'AFI 25 - 4.4	Undirected	332		2
49_MMG	AFSC_FM	{'AFI 25 - 4.4	Undirected	333		2
49_MMG	AF_A4LX	{'AFI 25 - 4.4	Undirected	334		2
49_MMG	AF_A4PY	{'AFI 25 - 4.4	Undirected	335		2
49_MMG	VEMSO	{'AFI 25 - 4.4	Undirected	336		2
49_MMG	WRM_GMO	{'AFI 25 - 2.3	Undirected	337		6

---

[2] We ran the Yifan Hu layout on the edge list, which places the most interconnected nodes centrally and the least interconnected nodes on the peripheries of a given graph.

For example, in the first line of Figure B.3, a relationship between the 49 MMG and the Air Force Civil Engineer Center (AFCEC) is established in AFI 25-101, Section 4.4. That relationship is shown in the visual representation in Figure B.4 (upper-left side). The section in the AFIs where the relationship is established (and where the detailed explanation can be referenced if needed) lies along the edge, or line, connecting these two entities (circled in blue). Figure B.4 further shows the connection between those two offices and the WRM global manager or SCOW (also referred to as the GMO [global management office] in Figure B.3 and other figures in this appendix), as well as other entities involved in the WRM process. Figure B.4 is part of the Set 1 analysis, which we turn to next.

Figure B.4. Example of Edge-List Conversion to Visual Representation

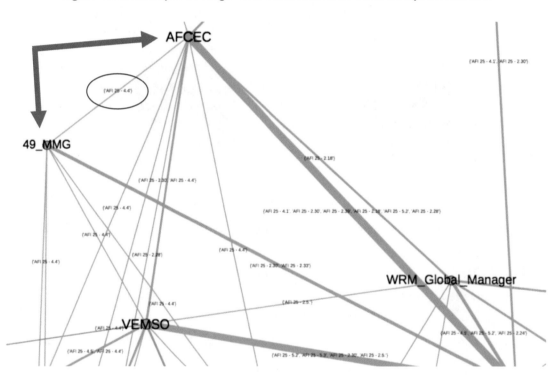

Set 1: Organizations/Offices/Working Groups to Organizations/Offices/ Working Groups

Figure B.4 is the upper-left corner of the larger Set 1 visualization shown in Figure B.5. The Set 1 visualization shows how Air Force WRM organizations, offices, and working groups interact with one another, using the AFI sections to establish those connections. Although the network analysis of Set 1 does not explain how those entities interact, it does show which entities interact most often. The organizations that work together most often are denoted by a color cluster. Some organizations may be connected to another organization in more than one respect; in those cases, more than one AFI is listed. Typically, where there is more than one link to an AFI, the edge between respective organizations is thicker. The nodes of organizations that act as hubs, provide pivotal functions, and interact with many other organizations are darker and larger than the nodes of organizations with peripheral functions. The WRM GMO or SCOW (the purple node) is one example.

Figure B.5. Set 1: Organizations/Offices/Working Groups to Organizations/Offices/Working Groups, Connected by AFI Sections

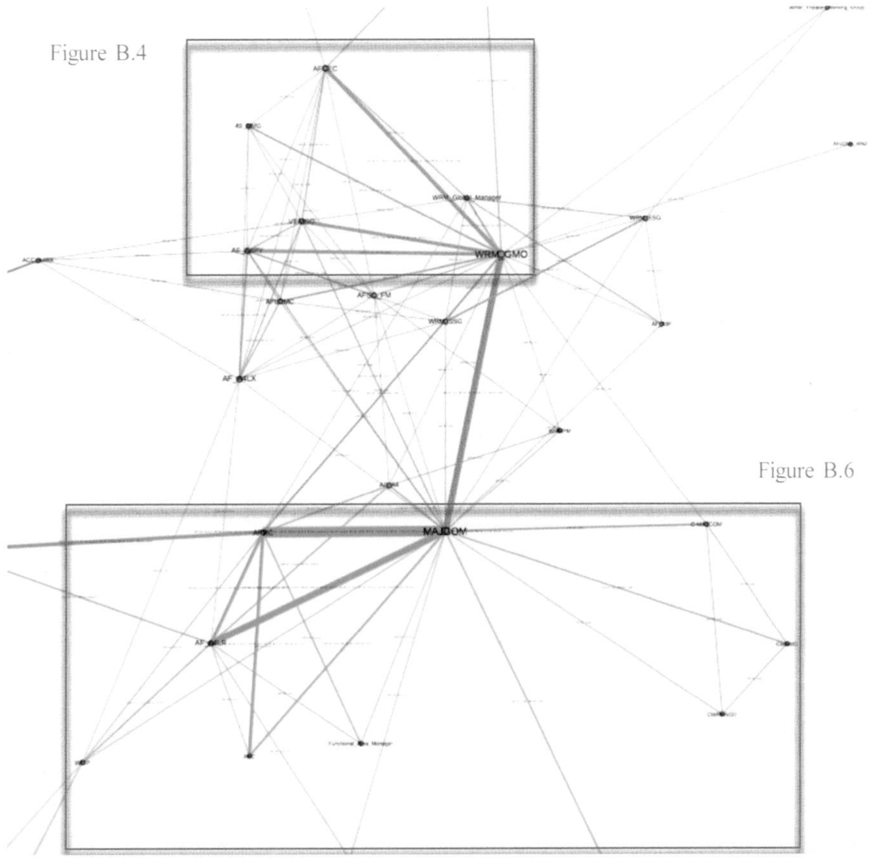

NOTE: Green represents the MAJCOM community; purple represents the WRM global manager community.

The lower section of Figure B.5 shows the relationships between a MAJCOM, denoted in green, and other WRM entities. A MAJCOM connects to other organizations, offices, and working groups, but those connections are denoted by a different color when the relationship lies outside the range of a particular community. In this case, MAJCOMs are more frequently connected to the activities of Air Force Materiel Command (AFMC), Air Mobility Command, AF/A4's Logistics Readiness Directorate (AF/A4LR), AF/A4's Resource Integration Directorate (AF/A4P), functional area managers, and other nodes of the same color. Figure B.6 shows a larger view of these connections.

71

Figure B.6. Set 1: Area of Detail for MAJCOM Clustered Group

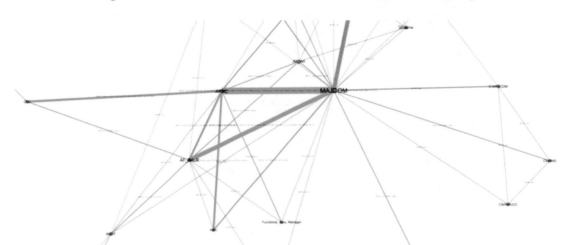

Set 2: Organizations/Offices/Working Groups to Systems/Tools/Guidance

The visualization for Set 2 is derived from a framework similar to that in Set 1, but it pairs organizations, offices, and working groups with systems, tools, and guidance. Through the research, SMEs identified several relevant WRM systems and tools—for example, fuels operational readiness capability equipment (FORCE) and BEAR, shown in Figure B.7. The visualization in the figure was helpful in showing the importance of these two WRM systems, but because of the multilayered overlap of clusters and groups, the visualization lacked the clarity we saw in Set 1, where clusters were more clearly grouped by color.

Figure B.7. Set 2: Organizations/Offices/Working Groups to Systems/Tools/Guidance, Connected by AFI Sections

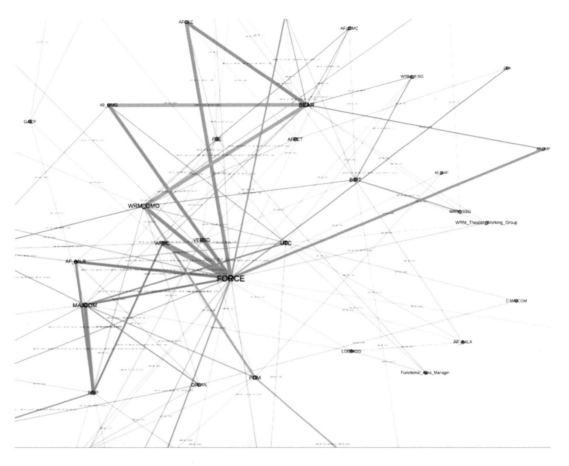

To provide a clearer view of how these systems interact, we highlighted only the first-order links attributed to each node by employing a simple function in Gephi that allows "hovering" over a particular node. For example, by hovering over FORCE, the lines of first-order connection are brought to the foreground, and all others are grayed out (see Figure B.8). Each colored line represents a community or an organization that is connected to FORCE in some way, with some frequency.

Figure B.8. First-Order Connections to FORCE

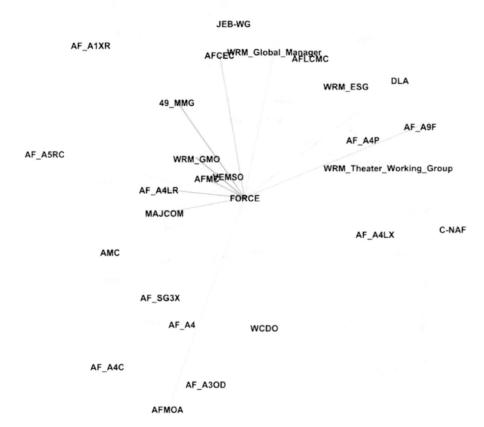

Set 3: Organizations/Offices/Working Groups to Organizations/Offices/ Working Group with Systems/Tools/Guidance

Figure B.9 is a high-level view of the connections between organizations, offices, and working groups through the varying systems, tools, and guidance. These include such pivotal systems and tools as FORCE, BEAR, the readiness spares package (RSP), time-phased force deployment data (TPFDD), and the Logistics Module (LOGMOD).

Figure B.9. Set 3: Organizations/Offices/Working Groups to Organizations/Offices/Working Groups, Connected by Systems/Tools/Guidance

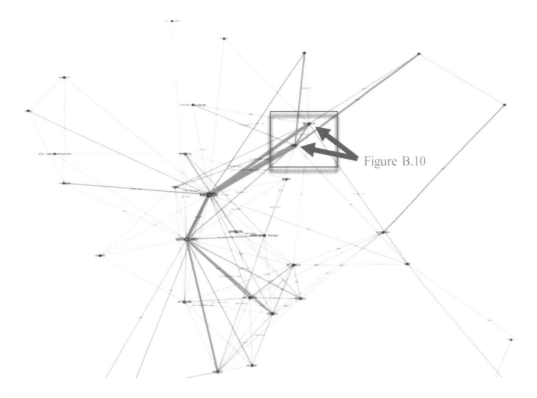

When we zoom into an area to see the detail, the name of the system, tool, or guidance is identified on the edge, or line, as seen in Figure B.10.

Figure B.10. Example of How a System, Tool, or Guidance Connects AFMC and AF/A4LR

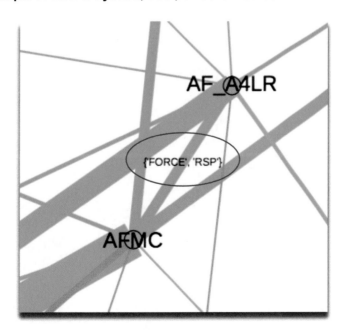

Set 4: Organizations/Offices/Working Groups to Verbs/Keywords

Set 4 initially produced a visualization that was difficult to follow because of the overlapping nodes and lines. Figure B.11 connects organizations, offices, and working groups to verbs and keywords, with AFI sections as the connecting factors. As the figure shows, certain verbs and keywords—such as, plan, manage, readiness, and equipment—are used more frequently than others. These frequently recurring terms are linked to the organizations most often associated with those verbs or key terms (for example, AFMC, MAJCOM, and WRM GMO), but the other connections are multidirectional and not easy to follow.

Figure B.11. Set 4: Organizations/Offices/Working Groups to Verbs/Keywords, Connected by AFI Sections

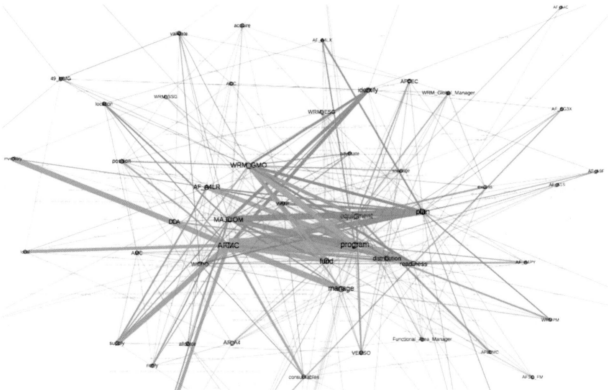

Set 5: Organizations/Offices/Working Groups to Organizations/Offices/ Working Groups with Verbs/Keywords

Figure B.11 was unhelpful when trying to determine how the organizations were related to each other through these verbs and keywords. As a result, we ran another visualization (Set 5) that repeated the framework established in Set 1—linking organizations, offices, and working groups to organizations, offices, and working groups—but rather than list the AFI references along the connecting lines, we listed the verbs and keywords to explain the connections between the organizations. The results are shown in Figure B.12, which resembles Figure B.9 but is distinct. By collapsing the matrix that originated Figure B.12, we could identify the relationships between organizations based on actions and things. The previous iteration, in Figure B.9, identified the connecting systems, tools, and guidance but did not reveal the functions that established the connection.

Figure B.12. Set 5: Organizations/Offices/Working Groups to Organizations/Offices/Working Groups, Connected by Verbs/Keywords

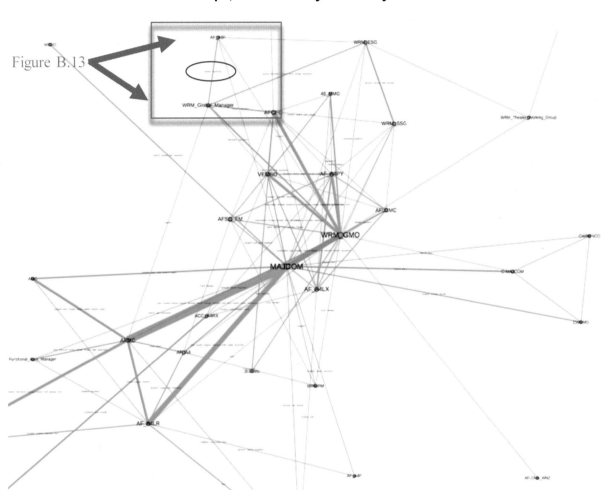

When we zoom into an area to see the detail, as in Figure B.13, we can see how this visualization identifies the actions (via verbs and keywords) that link two or more organizations together.

Figure B.13. Example of How a Verb or Keyword Connects the WRM Global Manager and AF/A9F

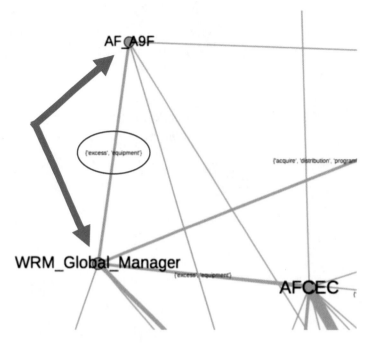

NOTE: AF/A9F = Air Force Deputy Chief of Staff for Studies, Analysis, and Assessments, Force Structure Analyses Directorate.

In this example, both AF/A9F and AFCEC are connected through the WRM global manager with the unique identifying keywords *excess* and *equipment*. Therefore, at first glance, the observer can clearly see that the WRM global manager plays some role with excess equipment, but to understand the relationship in greater detail, the analyst must reference the AFI section where the connection is established. To do so, the analyst can refer to the visualization produced in Set 1 (Figure B.5).

Challenges and Opportunities for Further Study

The network analyses explained here have varying levels of complexity and understandability. Some are simple to follow and show clear connections, while others are more difficult to understand. This is partly because of the oftentimes entangled layers of WRM processes. We use this information as a way to understand the complexities of the WRM processes as outlined in the AFIs.

Another use of this type of network analysis, which we did not employ here, would be to find the shortest distance between two organizations that do not have a first-order connection but that are connected through other relationships. This type of analysis could be helpful when seeking to identify redundancies, gaps, or inefficiencies in a particular process.

Analysis with the PRePO Model

This appendix describes the PRePO model and the process used to carry out the prepositioning analysis illustrated in Chapter Five.

Modeling Analysis for Global WRM Management

The intent of the analysis documented in this report is to illustrate that a modeling and analysis capability can be valuable to the global management of WRM. We demonstrate how modeling and analysis could help improve global effectiveness by quantifying the effects of various trade-off decisions. We illustrate how such a capability could be used to answer questions about, for example, how alternative regional storage postures affect readiness in a particular region and in other regions. Thus, the analysis described in this report presents notional results, demonstrating how modeling can help inform real-world decisions.[1] Application of this capability to a specific question would require input data specific to the system in question, and useful conclusions would require sensitivity analysis to clarify the effect of uncertainty. As a result, the analysis described in this report can offer insights into the relationship between storage postures and operational results, under certain conditions. But it does not provide answers to specific questions about prepositioning in specific regions, because it is not informed by the most current operational plans and because sufficient sensitivity analyses have not been conducted to overcome data shortcomings and inherent uncertainties.

Implementation of Global WRM Management Objectives

There are some differences in the PRePO model that we used for this analysis and the one used in previous RAND analyses.[2] To present analyses relevant to a global WRM manager, we made changes to enable the model to prioritize the transportation of some WRM types before others and to some locations before others. In short, the changes enable the model to consider timing differences across WRM types (for example, the WRM required for FOC can arrive after that required for IOC, as discussed in Chapter Five), bases (some FOLs are required to be operational before others), and theaters (time to set up and deploy are expected to be greater in some theaters than others). The timing capability enables the model to make meaningful trade-

---

[1] We have some unpublished slides that illustrate the costs and performance of the prepositioning COAs described in this appendix, applied to WRM stored for use in four different regional conflicts. These slides are not available to the general public but can be made available upon request.

[2] Thomas et al., forthcoming.

offs between positioning WRM close to the point of intended use and centralizing it at a more distant location. There may be financial advantages to storing part of a region's WRM requirement somewhere other than inside the region where it will be needed, especially when such positioning is done to create a shared pool of WRM among multiple regions, reducing the amount of stock required for purchase (although only one region could have use of the equipment from such a pool at any time) and enabling the sharing of maintenance personnel to sustain the WRM. Such an arrangement could also be of benefit when there is a limited supply of some equipment needed for multiple conflicts. However, storage outside the region of intended use imposes its own challenges, including increasing the need for inter-theater transport. Moreover, a shared pool of assets requires that WRM managers be capable of making allocation decisions quickly when WRM is needed, addressing the increased risk of shortfall in the event of another conflict, and recognizing differences in priority among WRM types when deciding which WRM to pool (and position afar). Pooling WRM at swing locations would require making trade-offs on the positioning of WRM storage sites on a global scale.[3]

To enable the PRePO model to support trade-off decisions on storage posture design and location and their operational effects, we modified the objective function of the model to assess the performance of storage postures, given WRM requirements made up of different priority categories. In the new PRePO model objective function, receiving the WRM required for IOC (also described as *closing* the WRM requirement for IOC) is now prioritized over receiving the WRM required for FOC, and the latter is prioritized over receiving WRM that will offset the loss of WRM through attrition. The objective function used for this analysis was as follows:

$$\text{Min } w = \alpha_1 \sum_{fol,wrm=IOC,t} folNotClosed[fol,wrm,t] * t^2 +$$
$$\alpha_2 \sum_{fol,wrm=FOC,t} folNotClosed[fol,wrm,t] * t^2 +$$
$$\alpha_3 \sum_{fol,wrm=ATTR,t} folNotClosed[fol,wrm,t] * t^2$$

The equation includes a new binary indicator variable, *folNotClosed*[*fol,wrm,t*], which we added to indicate WRM closure at an operating location for a specific WRM type *wrm*. The weights α determine the degree to which closing one type of WRM is prioritized over another. In general, $\alpha_1 \geq \alpha_2 \geq \alpha_3$, to encourage closing the IOC requirement over closing the FOC requirement and closing the FOC requirement over closing the requirement for offsetting attrition. The variable *folNotClosed*[*fol,wrm*] equals 1 if additional WRM of type *wrm* must be transported to the location *fol* at time *t* (because it has not yet been transported and was not stored collocated at the operating location).

---

[3] Shared stocks are needed to support the 2018 National Defense Strategy objectives. Joint and Air Force doctrine both support the idea of swing stock to support more than one AOR.

Modeling Assumptions

We made a few simplifying assumptions for this analysis, based on input from RAND SMEs, and we discuss these assumptions in the following sections.

Theaters and Scenarios

The analysis was conducted across three theaters and included four scenarios. We used the 2018 National Defense Strategy to prioritize the conflicts.[4] The results shown in this report are unclassified and notional.

Forward Operating Locations and WRM Requirements

The FOLs used in the analysis represent operational locations consistent with operations plan beddowns for each of the four conflicts modeled. The WRM requirements for these locations were derived from Lean-START for all locations with aircraft squadrons in the beddowns, taking into account different levels of capability already existing at each location (with more or less WRM required to supplement existing capabilities, according to Lean-START output).

Consistent with guidance in the 2018 National Defense Strategy, we assumed that WRM was authorized and prepositioned in preparation for two of the four conflicts modeled (referred to in this report as regions A and B). FOLs used to prosecute the other two conflicts (located in regions C and D) were assumed to have to rely on WRM allocated for conflicts in regions A and B (assuming that any WRM that might have been allocated to these regions previously has become obsolescent or been removed).

WRM Storage

WRM storage was assumed to occur in regions A and B (including on the boundary of these regions), except when allocated to swing locations (assumed to be in CONUS). The WRM required for offsetting attrition was always stored at swing locations. The WRM required for IOC and FOC was stored under one of the following COAs:

- Highly distributed regionally: WRM is assumed to be stored at locations near enough to the operating bases to be functionally collocated (this may be on-base storage or adjacent commercial storage).
- Moderately distributed regionally: WRM is assumed to be stored at six locations, dispersed throughout the region, based on FOLs. Using six locations enables the full use of all regional lift available, according to modeling assumptions, and enables WRM to be well distributed throughout the theaters in which it is prepositioned.
- Centralized regionally: WRM is assumed to be stored at two sites located in the center of the region (not necessarily located near different FOLs).

---

[4] Thomas et al., forthcoming.

- Centralized at swing locations: WRM is assumed to be stored in CONUS and ready for immediate shipment to a region with a requirement.

For example, under these storage COAs, the WRM required for IOC (IOC WRM) may be highly distributed regionally, whereas the WRM required for FOC (FOC WRM) is centralized at swing locations. Only a subset of the possible storage COA combinations were considered in the modeling. Those COAs are as follows:

1. IOC WRM highly distributed, FOC WRM centralized regionally
2. IOC WRM highly distributed, FOC WRM centralized at swing locations
3. IOC WRM moderately distributed, FOC WRM moderately distributed
4. IOC WRM moderately distributed, FOC WRM centralized at swing locations
5. IOC WRM centralized regionally, FOC WRM centralized regionally
6. IOC WRM centralized regionally, FOC WRM centralized at swing locations.

The swing COA is used for storing the WRM required to offset attrition and sometimes that required for FOC. In this analysis, when region A or B uses a COA that assigns the WRM required for FOC to swing locations, it is because the total global supply of that WRM type is not enough to meet the total requirement for both regions. Thus, if either region A or B uses a swing COA, both are using a swing COA. The swing COA implies that the smallest quantity of WRM required to meet both regions' requirements has been acquired (the largest of the two regions' requirements). This has implications both for the purchasing cost of the global WRM inventory and for operational risk.

It should be noted that, because WRM is assumed to be allocated to only two of the conflicts, regions C and D must rely on storage sites in regions A and B. And the storage configurations used in regions A and B could affect performance in regions C and D. Therefore, when theaters C and D are modeled, the storage COAs used apply to storage in regions A and B (not within regions C and D). In order to simplify the modeling, for a specific model run, storage COAs in regions A and B are assumed to be identical. In other words, when modeling region C under COA 1 in the list above, COA 1 is assumed to apply to both regions A and B; no combinations of the regional storage COAs were modeled. However, because regions C and D are never supplied by more than one region in any model run, this simplification does not affect the results.

In addition, regions C and D were never examined under COAs 1 and 2. This is because, when using only strategic lift resources or using strategic lift resources plus intra-theater airlift, COAs 1 and 2 are indistinguishable from COAs 3 and 4, respectively, because of transportation asset limitations (this is true for all assumed values about transportation availability).

Transportation

In the PRePO model, several transportation modes—including land, air, and sea cargo vehicles, as well as intra-theater and inter-theater (strategic) lift—can be used to transport WRM from storage location to demand sites. We evaluated a range of values for the number of vehicles that might be available to support WRM movement for each mode type. This analysis considered two alternatives: the availability of transportation assets is high or such availability is low, as presented in Table C.1.

Table C.1. Availability of Transportation Assets

	Low	High
Strategic Airlift (in C-17 equivalents)	2	4
Intra-theater airlift (in C-130 equivalents)	35	76
Trucks	100	300
LMS-ROROs	2	2
HSCATs	2	2

NOTE: HSCAT = high-speed catamaran; LMS-RORO = large, medium-speed roll-on, roll-off.

In the PRePO model, no constraint is placed on the starting locations of vehicles; rather, the model selects the initial location of each transportation asset. Each vehicle has a maximum capacity, a travel speed, and a cost per ton-mile of use (assuming full cargo loads).

Throughput at Storage Facilities, Forward Operating Locations, and Airports

Storage facilities, FOLs, and airports have maximum throughput capacities. We assumed that storage depots are capable of serving 50 trucks per day, FOLs are able to receive 24 trucks per day, and airports at FOLs and FSLs are capable of handling up to 16 C-130s per day or 12 C-17s.

Freedom of Movement and Border Crossings

In some of the modeled regions, there are restrictions on the movement of cargo across political boundaries. In such regions, prior approval of diplomatic clearance may be required for travel among multiple countries. This includes multi-country itineraries for overflight or landing of cargo transported by aircraft, as well as for multi-country overland shipping routes. However, the nature of diplomatic clearance requirements during the setup phase of a conflict is uncertain. It is possible that, for at least some countries and shipments, arrangements for waiving or expediting diplomatic clearances could be made before transport of WRM begins. In this analysis, we do not consider the potential delays generated by diplomatic clearance requirements and assume that all such requirements have been met or waived once the in-theater movement of WRM has begun.

Customs inspections can also create delays. This analysis assumes that trucking shipments are delayed by 0.5 days, on average, for overland shipments between distributed storage sites and operational locations; such shipments are delayed one full day for shipments between centralized storage sites and operational locations; and no other transportation mode suffers border crossing delays. These penalties are applied uniformly to all trucking shipments meeting the conditions. Calculating the number of borders crossed for each itinerary is complicated and not necessarily constant between two locations when more than one route is possible. This assumption could have the effect of overestimating or underestimating by a few days the time it takes individual bases to receive the required WRM.

Additional Modeling Results

Interpreting the Results Tables

Tables C.2 through C.4 show the aggregated results for regions A, B, C, and D. The first column in each table indicates the storage COA used. The notation *distributed* means moderately distributed (at six storage sites dispersed throughout the region), whereas *at FOLs* means that WRM was highly distributed (collocated at FOLs for practical purposes, either at the base or at commercial sites adjacent to the base).

For regions A and B, the cost values presented reflect the one-time purchase cost of WRM and the annual cost of storage and maintenance for the WRM required for that region (for a specific conflict) using the COA listed. The storage and maintenance costs include annual facility costs as variable costs of storage (per square foot of warehouse space) and the cost of maintenance and obsolescence (assumed to be 5 percent of the WRM purchase cost, annually). For regions C and D, no WRM is purchased specifically to support these regions, and none is stored or maintained in these regions; therefore, none of the costs shown are specific to the region or its requirements. Instead, they are global costs to purchase, store, and maintain a WRM inventory globally. That global WRM inventory and posture have direct effects on the ability to agilely support conflicts in regions C and D, so the costs of supporting each COA are shown. These costs are simply the sum of the costs to purchase, store, and maintain WRM in regions A and B.

As stated earlier, the cost to purchase WRM varies depending on whether the WRM required for FOC is assigned to a swing location. When a swing COA is used, it implies that only a subset of the total WRM requirement for regions A and B has been purchased (the minimum required to support a single conflict in either region). Instead, when a non-swing COA is used, it is assumed that the full WRM requirement for FOC for both theaters has been purchased.

Table C.2. Region A Cost and WRM Closure Time

WRM Storage COA		Cost (in $Millions)		IOC WRM Closure (Days)		FOC WRM Closure (Days)		Full WRM Closure (Days)	
IOC	FOC	Purchasing	Storage & Maint.	High Tran.	Low Tran.	High Tran.	Low Tran.	High Tran.	Low Tran.
Centralized	Swing	448	25	13	18	24	26	25	27
Distributed	Swing	448	27	5	13	24	26	25	27
at FOLs	Swing	448	33	0	0	24	25	25	26
Centralized	Centralized	491	28	13	20	16	22	20	23
Distributed	Distributed	491	29	5	13	7	15	19	21
at FOLs	Centralized	491	35	0	0	4	7	20	20

Table C.3. Region B Cost and WRM Closure Time

WRM Storage COA		Cost (in $Millions)		IOC WRM Closure (Days)		FOC WRM Closure (Days)		Full WRM Closure (Days)	
IOC	FOC	Purchasing	Storage & Maint.	High Tran.	Low Tran.	High Tran.	Low Tran.	High Tran.	Low Tran.
Centralized	Swing	276	16	20	25	30	31	30	34
Distributed	Swing	276	17	11	16	29	31	29	33
at FOLs	Swing	276	23	0	0	29	31	29	33
Centralized	Centralized	324	18	20	25	26	32	26	32
Distributed	Distributed	324	20	11	16	16	20	26	32
at FOLs	Centralized	324	25	0	0	9	13	23	29

The results for regions A and B follow expected patterns. In the case of IOC WRM, we see that the more distributed the storage posture, the faster the region's WRM requirements are closed. The same is true for FOC WRM and FOC storage posture. Moreover, FOC WRM closure performance is also affected by the IOC posture used. This is not surprising, because the earlier that IOC WRM requirements are closed, the earlier that lift assets can start making progress on transport of FOC WRM. The exception is when a swing COA is used for FOC WRM. In that case, closing IOC WRM requirements early has very little effect on closing FOC requirements, because regional transportation assets cannot be levied for FOC WRM until ships have arrived from swing locations in CONUS with FOC WRM cargo. The closure performance results for WRM required to offset attrition follow a similar pattern.

The cost of each COA is broken into purchasing cost and annual cost.[5] The purchasing cost of each option increases when all FOC WRM is purchased (as opposed to purchasing a smaller quantity of FOC WRM, consolidated at swing locations, enough for only the largest of the regional FOC WRM requirements). The annual investment required depends on the amount of WRM purchased, as well as the storage COA. Greater dispersal of storage requires more storage sites and higher fixed costs related to storage facilities. These differences are small compared with the purchasing costs of WRM for shorter periods of storage time.

---

[5] We use these costs as an illustration of the type of analysis that should be completed when evaluating storage postures. Life-cycle costs should also be assessed. Including life-cycle costs would help decisionmakers understand the relative total cost of each option. We did not include life-cycle costs in this notional example because we did not have the data to do so.

The results tables for regions C and D (Tables C.4 and C.5) show similar patterns to those of regions A and B in that dispersion of storage and increases to transportation asset availability improve closure time results. However, both distribution of storage and increasing transportation assets have much less effect on WRM closure for these regions. Also, region 4 shows worse overall closure performance than region 3 does, illustrating that differences in geography and total WRM requirements can outweigh the performance advantages granted by storage posture and transport availability, in some cases.

There is no cost to support regions C and D, specifically, because they rely only on the WRM that is purchased and stored for the other two regions. However, the total global purchasing and annual costs are shown for context for each COA, because the COAs do affect WRM closure at bases in regions C and D.

Table C.4. Region C Cost and WRM Closure Time

| WRM Storage COA | | Cost (in $Millions) | | IOC WRM Closure (Days) | | FOC WRM Closure (Days) | | Full WRM Closure (Days) | |
IOC	FOC	Purchasing	Storage & Maint.	High Tran.	Low Tran.	High Tran.	Low Tran.	High Tran.	Low Tran.
Centralized	Swing	724	41	22	23	25	25	26	26
Distributed	Swing	724	43	20	21	22	22	25	25
Centralized	Centralized	814	46	22	23	25	25	26	26
Distributed	Distributed	814	48	20	21	20	22	25	25

Table C.5. Region D Cost and WRM Closure Time

| WRM Storage COA | | Cost (in $Millions) | | IOC WRM Closure (Days) | | FOC WRM Closure (Days) | | Full WRM Closure (Days) | |
IOC	FOC	Purchasing	Storage & Maint.	High Tran.	Low Tran.	High Tran.	Low Tran.	High Tran.	Low Tran.
Centralized	Swing	724	41	13	13	22	23	29	29
Distributed	Swing	724	43	10	15	20	23	27	29
Centralized	Centralized	814	46	12	16	14	17	23	29
Distributed	Distributed	814	48	10	16	10	16	22	29

SPRAT Data Sources

SPRAT provides an overview of 15 qualitative political, economic, and strategic characteristics across 195 countries to assist A4 planners in evaluating the willingness, capacity, and reliability of potential host nations to safely secure, store, and provide access to prepositioned WRM assets within their territories. (For a summary of this tool and an illustration of its use, see Chapter Four.) This appendix provides a detailed description of the sources, methods, and coding schemes used to populate each of the 15 measures in SPRAT's Microsoft Excel–based framework, as well as suggestions for planners on how to interpret and weigh each of these considerations based on the reliability of the data and the degree to which they serve as adequate proxies for the characteristics they intend to measure.

SPRAT Color-Coding Scheme

In addition to the binary (yes/no), ordinal (none/low/medium/high), and categorical (none/restricted/denied) data values that populate each value cell in SPRAT, each value cell was assigned a color according to the severity that each data value represents relative to the data values assigned to the other countries in the same measure column. Green indicates that a country has characteristics one would ideally desire a host nation to have for the particular measure in question, representing a relatively low level of concern or risk when weighing WRM prepositioning. Yellow or orange indicates that a country has less favorable or desirable characteristics for the measure in question, representing a relatively moderate level of concern or risk when weighing WRM prepositioning. Red indicates that a country has unfavorable or undesirable characteristics for the measure in question, representing a relatively high degree of concern or risk in weighing WRM prepositioning. Gray indicates either that the value of the measure in question reflects neither favorably nor unfavorably on a country and should be interpreted neutrally or that insufficient data were available to make an accurate assessment of the country. This color scheme is summarized in Table D.1.

Table D.1. SPRAT Color-Coding Scheme

Level of Risk	Favorability of Value	Value Cell Color Coding
Low	Favorable	Green
Medium	Less Favorable	Yellow
Moderate	Less Favorable	Orange
High	Unfavorable	Red
Neutral	Neither	Gray

The color-coding scheme was devised to provide planners with a quick general overview of the favorability of a candidate country for the prepositioning of WRM assets by allowing users to quickly scan across rows to get a general feel for a candidate country based on an ordinal, thermometer-like scale. Countries whose values include more green cells across the 14 categories measured are more likely to be favorable WRM prepositioning locations, while countries with more yellow, orange, or red cells are more likely to be less favorable WRM prepositioning locations based on the country's estimated willingness, capability, and reliability in securing, storing, and providing access to WRM assets.

Host-Nation Capacity and Capability to Secure and Facilitate the Movement of Prepositioned WRM

In one general category, SPRAT supplies planners with information about the capacity of a candidate host nation to securely and reliably store prepositioned WRM assets for use in peacetime and contingency operations (see Table 5.1).

Internal Stability of the Host Nation

Information about the internal stability of a host nation's government is drawn from the global country risk ratings published on August 14, 2018, by Jane's Military and Security Assessments Intelligence Centre from IHS Markit. Analysts from Jane's Economics and Country Risk team assign scores to each of 196 countries across a host of metrics that aggregate into overall scores in five broad categories: political risk, external risk, internal security risk, infrastructure risk, and economic risk. These scores range from 0.1 to 10, and steps of 0.1 on a logarithmic scale reflect the average expected level of risk in each of these domains over the next year. These scores are based on IHS Markit economists' and country risk analysts' understanding of each country's political, economic, social, and security environment using economic models, information from open sources, and structured intelligence-gathering by a network of thousands of in-country personnel.[1]

A country's political risk score is an average of four sub-scores that reflect risks of

- government instability, including the degree to which ruling individuals or a centralized government extends sovereign control over territory or its population and the risk that a country's regime will be forced to change by irregular succession pathways, such as coups, popular uprisings, votes of no confidence, impeachment, and civil war
- policy direction shifts, including the risk that the government's broad policy and regulatory framework changes over the next year in ways that could adversely affect international commerce, such as the imposition of tariffs or quotas or major shifts in employment and environmental regulations
- state failure, including the degree to which a country wields an effective monopoly over the use of force to ensure public safety, law and order, and the provision of basic goods and services (such as food, sanitary water, infrastructure, and energy)
- bribery and corruption, including the likelihood that companies will confront demands for bribes or other corrupt practices in order to do business.

Within each of these four sub-categories, sub-scores range from 0.1 to 10, with higher values reflecting higher degrees of political risk. For government instability, for example, scores ranging from 0.1 to 2.3 indicate governments that are well established and internally stable, as well as those that are stable or fairly stable but may experience occasional challenges to the leadership in the ordinary course of politics and through regular institutional channels. Scores ranging from 2.4 to 4.3 indicate fairly weak coalition governments that face the prospect of dissolution from an increasingly organized opposition, as well as governments that face frequent, widespread, and increasingly bold challenges to their authority from civil society and political opposition groups. Scores of 4.4 or higher indicate governments whose authority is severely challenged by domestic and international actors who are likely to change the regime through irregular or forceful means, as well as governments that are internally unstable and are severely challenged by violent uprisings, insurgencies, or other actors who engage in civil war. The sub-scores for policy direction, state failure, and bribes and corruption fall along a similar scale, and the sub-scores for each of these four factors are averaged to provide an aggregate political risk score from 0.1 to 10, with higher scores reflecting higher political risks.

SPRAT inverts this scale so that, when a country received an aggregate political risk score of 4.4 or above, it was coded as "low" for internal stability. Countries receiving an aggregate political risk score of between 2.4 and 4.3 were coded as "medium," and countries receiving an aggregate political risk score of 2.3 or below were coded as "high."

---

[1] Jane's Military and Security Assessments Intelligence Centre, 2018.

Internal Security of the Host Nation

Information about the internal security of a host nation's government is drawn from Jane's global country risk ratings published on August 14, 2018, by Jane's Military and Security Assessments Intelligence Centre from IHS Markit. This same data source was used to construct measure 9.1.6 of the RAND Security Cooperation Prioritization and Propensity Matching Tool. See the earlier discussion about how analysts from Jane's Economics and Country Risk team assign scores to countries in five broad categories.[2]

A country's internal security risk score is an average of six sub-scores that reflect a country's susceptibility to internal conflict (including the likelihood and effect of civil military conflict among governments, insurgents, separatists, or other armed actors); terrorism and other acts of violence by organized non-state actors; kidnapping and demands for ransom; major protests and riots; and the likelihood of injury or death from organized or unorganized acts of violence.

Within each of these six sub-categories, sub-scores range from 0.1 to 10, with higher values reflecting higher degrees of internal security risk. For internal conflict, for example, scores ranging from 0.1 to 2.3 indicate countries that lack the kinds of grievances that would motivate armed resistance, as well as those that do have historical grievances that either are aired only occasionally or are expressed through normal institutional channels or isolated incidents. Scores ranging from 2.4 to 4.3 indicate countries that experience either (1) a limited insurgency fought against state security forces and in which fighting is contained mostly to rural areas or (2) a more serious, sustained campaign in which there is some damage to infrastructure and population centers, and insurgents occasionally are able to hold certain defensible positions. Scores of 4.4 or higher indicate countries experiencing or at imminent risk of a full-fledged civil war, with state military forces engaged in operations throughout the country against well-armed, and sometimes foreign-backed, insurgent forces. The sub-scores for terrorism, kidnapping and ransom, protests and riots, and the likelihood of injury or death fall along a similar scale, and the sub-scores for each of these six factors are averaged to provide an aggregate internal security risk score ranging from 0.1 to 10, with higher scores reflecting higher internal security risks.

SPRAT inverts this scale so that, when a country received an aggregate internal security risk score of 4.4 or above, it was coded as "low" for internal security. Countries receiving an aggregate internal security risk score of between 2.4 and 4.3 were coded as "medium," and countries receiving an aggregate internal security risk score of 2.3 or below were coded as "high."

External Security of the Host Nation

Information about the external security of a host nation's government are drawn from Jane's Global Country Risk Ratings published on August 14, 2018, by Jane's Military and Security

---

[2] Jane's Military and Security Assessments Intelligence Centre, 2018.

Assessments Intelligence Centre from IHS Markit. This same data source was used to construct measure 9.1.7 of the RAND Security Cooperation Prioritization and Propensity Matching Tool. See the earlier discussion about how analysts from Jane's Economics and Country Risk team assign scores to countries in five broad categories.[3]

A country's external security risk score ranges from 0.1 to 10, with higher values reflecting a higher likelihood that the country would be engaged in inter-state conflict. Scores ranging from 0.1 to 2.3 indicate countries that lack territorial disputes or other serious conflicts of interest with their neighbors and that present no more than a moderate risk of violent conflict, as well as countries whose poor relations with some neighboring countries lead to occasional border closures, cross-border fire exchanges, or other isolated violent incidents that do not escalate into full-blown conflict. Scores ranging from 2.4 to 4.3 indicate countries that maintain poor relations with neighboring countries that result in regular and disruptive closures, seizures, or violent incidents, as well as countries that are engaged in long-standing territorial or other disputes and where armed forces are maintained at high levels of combat readiness and are involved in frequent cross-border clashes involving artillery and other heavy weapon fire, air strikes, or ground incursions. Scores of 4.4 or higher indicate countries experiencing or at imminent risk of a full-fledged inter-state war, and opposing military forces are engaged in major combat operations.

SPRAT inverts this scale so that countries were coded as "low" for external security if they received an external security risk score of 4.4 or above. Countries were coded as "medium" if they received an external security risk score of between 2.4 and 4.3, and countries were coded as "high" if they received an external security risk score of 2.3 or below.

Host-Nation Infrastructure Quality

Assessments of a country's overall infrastructure quality were derived from the World Bank's 2018 Logistics Performance Index (LPI), an interactive benchmarking tool that rates the quality of a country's infrastructure based on a worldwide survey of logistics operators (global freight forwarders and express careers) who provide feedback on six components of a country's logistics sector performance.[4] These components are the efficiency of customs and border clearances, the quality of trade and transportation infrastructure, the ease of arranging competitively priced shipments, the competence and quality of logistics services, the ability to track and trace consignments, and the frequency with which shipments reach consignees within scheduled or expected delivery times. Survey respondents are asked to rate the quality, efficiency, or ease of each of these components on a five-point scale ranging from 1 ("very low")

---

[3] Jane's Military and Security Assessments Intelligence Centre, 2018.

[4] Jean-Francois Arvis, Lauri Ojala, Christina Wiederer, Ben Shepherd, Anasuya Raj, Karlygash Bairabayeva, and Toumas Kiiski, *Connecting to Compete 2018: Trade Logistics in the Global Economy*, Washington, D.C.: International Bank for Reconstruction and Development/World Bank, 2018.

to 5 ("very high"). Responses are then aggregated across all six components into an overall LPI score ranging from 1 to 5 for each of the 160 countries in the data set.

For SPRAT, a potential host nation was coded as having "low" infrastructure quality if it received an overall LPI score of 2.5 or below. A country was coded as having "medium" infrastructure quality if it received an LPI score between 2.51 and 3.99, and a country was coded as having "high" infrastructure quality if it received an LPI score of 4.0 or above.

Host-Nation Climate Conditions

Information about a host nation's climate conditions comes from the 2004 Environmental Vulnerability Index (EVI), a multiyear effort by the South Pacific Applied Geoscience Commission to estimate each country's vulnerability to environmental damage and degradation based on 50 climate and environmental hazard indicators, such as sea temperatures, habitat loss, degradation rates, and frequency and intensity of natural disasters (such as earthquakes, tsunamis, and volcanic eruptions).[5]

For each of the 50 indicators, the study authors assigned countries an EVI score that ranged from 1 (low vulnerability/high resiliency) to 7 (extremely high vulnerability). They then aggregated these scores to produce a single overall EVI score for each of the 235 countries and territories in the study, and those scores ranged from 174 (low vulnerability) to 450 (high vulnerability). Countries were then classified into one of five vulnerability groups according to their overall EVI scores:

- resilient: overall EVI score of 215 or below
- at risk: overall EVI score from above 215 to 265
- vulnerable: overall EVI score from above 265 to 315
- highly vulnerable: overall EVI score from above 315 to 365
- extremely vulnerable: overall EVI score above 365.

In SPRAT, countries were coded as having "low" vulnerability to major climate events and disruptions if they were grouped into the "resilient" category based on their overall EVI scores. Countries were coded as having "medium" vulnerability to major climate events and disruptions if they were grouped into the "at risk" or "vulnerable" categories. Countries were coded as having "high" vulnerability to major climate events and disruptions if they were grouped into the "highly vulnerable" or "extremely vulnerable" categories.

Host-Nation Willingness and Reliability in Storing and Granting Access to Prepositioned WRM

In another general category of SPRAT measures, planners are supplied with information that we expected would shape a host country's willingness to store and grant access to prepositioned

---

[5] Kaly, Pratt, and Mitchell, 2004.

WRM assets (see Table 5.1). The first set of characteristics in this category are designed to capture elements of the nation's overall political and security relationship with the United States. The second set of characteristics in this category aim to capture a candidate host nation's political, economic, and security relationship with Russia and China, the United States' near-peer competitors. A third set of characteristics aims to provide planners with a summary of the candidate country's experience with hosting U.S. forces and any restrictions or denials of access it may have imposed on the United States historically during peacetime and contingency operations. In the following sections, we describe each measure associated with these three sub-categories.

Existing WRM Storage

Countries were identified as currently storing WRM assets based on an examination of a spreadsheet listing non-munitions WRM equipment with the countries' associated MAJCOM codes; the spreadsheet was provided by the sponsor and was current as of April 10, 2018. In SPRAT, countries were coded as "yes" if they stored at least one piece of equipment in the April 2018 data pull and "no" otherwise.

Existing Security Agreement

Membership in mutual defense treaties was derived from a comprehensive database of U.S. security agreements compiled as part of a RAND analysis in 2014.[6] The database identifies and classifies bilateral and multilateral treaties between the United States and other countries that were in force any time between 1955 and 2012. These same data were used to generate Measure 3.3.4 of the RAND Security Cooperation Prioritization and Propensity Matching Tool. We then updated the data on mutual defense treaties data to reflect the ascension of Montenegro into NATO in July 2017 and the announcement of the intentions of Bolivia, Ecuador, Nicaragua, and Venezuela to withdraw from the Inter-American Treaty of Reciprocal Assistance (Rio Treaty) in June 2012.[7]

A country was coded as "yes" if the host nation remains party to a bilateral or multilateral security agreement that provides mutual defense guarantees, such as the North Atlantic Treaty; the Inter-American Treaty; and the Australian, New Zealand, United States Security Treaty.

---

[6] Kavanagh, 2014. The data encompass 3,223 bilateral treaties signed by the United States and another partner. Multilateral treaties are disaggregated into 2,325 bilateral pairs between the United States and a single multilateral signatory. This produces 5,548 individual entries in the database.

[7] It should be noted that agreements are considered lapsed or withdrawn in Kavanagh's (2014) database only when formal legal procedures are undertaken to effectuate these plans, so the withdrawal of the four states from the Rio Treaty would have been reflected in the database only once formal action was taken to do so. On Montenegro's accession, see NATO, "Montenegro Joins NATO as 29th Ally," press release, June 9, 2017b; on the Rio Treaty, see James Bolsworth, "Cold War Defense Treaty Under Fire in Latin America," *Christian Science Monitor*, June 8, 2012.

A country was coded as "no" if it lacks such an agreement, even if it maintains some kind of security agreement and cooperation with the United States outside of these two designations.[8] We chose to exclude treaties and security agreements that bear on other important aspects of the country's overall security relationship with the United States, including agreements outlining the terms of joint military operations, access agreements, grants of financial assistance, equipment sales and transfers, and the administrative and legal provisions associated with the implementation of such sales and transfers.

Although these agreements can reveal important aspects of the security relationship with the United States, many are tied to particular historical contingencies that may be less useful for planners making forward-looking decisions on the prepositioning of global WRM. To the extent that a host nation's history of providing access and military cooperation to the United States is useful, this information has been supplied in other categories (such as existing WRM storage sites and peacetime and contingency access challenges) in ways that bear more directly on the choices that planners have to make about WRM prepositioning.

Participation in Major U.S.-Led Coalition Operations

This category measures whether a candidate host nation meaningfully contributed to major U.S.-led coalition operations since September 11, 2001, including those in Afghanistan, Iraq, and Libya, and in the global coalition to counter the Islamic State. Although other countries may have made contributions alongside U.S. forces in other operations, such as humanitarian assistance and disaster response missions, we believed that participation in the major, high-priority missions for the United States would provide the most reliable indicator of a host nation's willingness and ability to support U.S. forces in major contingency operations.[9]

Surprisingly, no single repository of data exists to identify the scope and scale of individual country contributions to each of these coalition operations, partly because different government departments and agencies maintain different measures for what counts as contributions to each of these operations. According to a U.S. Army–issued monograph on allied participation in Operation Iraqi Freedom, the number of foreign nations participating in the U.S.-led coalition is unknown; the White House, the State Department, and U.S. Central Command all maintain slightly different lists of countries that provided direct and indirect support to the operation. Whereas the Army study determined that approximately 60 countries provided direct or indirect assistance in the form of basing rights, commercial shipping, overflight, and humanitarian aid,

---

[8] Although the mutual defense agreement establishing the Southeast Asia Treaty Organization (SEATO) was not formally rescinded, it effectively abrogated with the disbanding of the SEATO organization itself in June 1977. SPRAT therefore excludes membership in SEATO from the analysis.

[9] We focused on the four most recent major military operations. The tool could be expanded to include other major military operations, such as Operation Desert Storm in 1991 and Operation Allied Force in 1999.

the study identified 37 nations that furnished direct support in the form of troop deployments for the operation from March 2003 until July 2009.[10]

Countries were counted as having participated in U.S.-led coalition operations in Iraq and Afghanistan if they made formal force deployments to these efforts. Countries were counted as having participated in the coalitions against Libya and the Islamic State if they made any kind of formal contribution to the effort, such as formal force deployments, financial contributions, and enhancements to political consultations. As noted in Chapter Five, information regarding foreign partner contributions in support of NATO's International Security Assistance Force and Resolute Support Mission in Afghanistan was derived from periodic summary reports of foreign troop deployments archived on NATO's website.[11] Information regarding foreign troop contributions to Operation Iraqi Freedom was derived from the aforementioned 2011 Army monograph.[12] Information regarding country contributions to the 2011 intervention in Libya was derived from a *Foreign Affairs* article coauthored by former U.S. Ambassador to NATO Ivo H. Daalder and former Supreme Allied Commander Europe James G. Stavridis.[13] Country participation in the 79-member Global Coalition Against Daesh was derived from the list of partners on the coalition's website.[14]

For SPRAT, countries were coded as "all" if they participated in all four major coalition operations. Countries were coded as "some" if they participated in at least one of the major coalition operations since September 11, 2001. Countries were coded as "none" if they did not participate in any of the four major coalition operations.

Host-Nation Political and Economic Cooperation with the United States

To capture the overall nature of political and economic cooperation between candidate host nations and the United States, SPRAT draws on measures and constructs that constitute category 3 in the RAND Security Cooperation Prioritization and Propensity Matching Tool. Category 3 provides an assessment of the long-term relationship of the partner nation and the United States across a variety of political and economic domains.[15] Our analysis draws on the data displayed in category 3 of the 2017 version of the Propensity Matching Tool spreadsheet, and we updated data on the volume of trade between the United States and the potential host nation as a percentage of GDP,[16] the flow of foreign direct investment between the United States

---

[10] Carney, 2011.

[11] NATO, 2017a.

[12] Carney, 2011.

[13] Daalder and Stavridis, 2012.

[14] Global Coalition Against Daesh, undated. The U.S. government typically refers to Daesh (an Arabic-language acronym) as the Islamic State of Iraq and Syria, or ISIS.

[15] For a complete list of measures and constructs that constitute this category, see Paul et al., 2013, Appendix A.

[16] U.S. Census Bureau, "U.S. Trade in Goods by Country," webpage, undated.

and the candidate host nation,[17] the percentage of a potential host nation's population that approves of U.S. leadership,[18] and the degree to which a potential host nation's voting behavior coincided with that of the United States at the United Nations General Assembly.[19] We then aggregated these measures and constructs using an adjusted weighting scheme to generate an overall propensity score for that category ranging from 0 to 1; higher values reflect a long-term relationship with more shared interests and greater political and economic cooperation.

In SPRAT, countries receiving a propensity score of 0.33 or below for category 3 were coded as having a "low" degree of overall political and economic cooperation with the United States. Countries receiving a propensity score between 0.34 and 0.66 for category 3 were coded as having a "medium" degree of overall political and economic cooperation with the United States, and countries receiving a propensity score of 0.67 or above were coded as having a "high" degree of overall political and economic cooperation with the United States.

Host-Nation Security Cooperation with Russia and China

To identify possible sources of leverage that near-peer competitors might exact over potential host nations, SPRAT seeks to measure the percentage of arms shipments to a host nation that come from Russia and China. For this measure, SPRAT draws on global arms transfers data collected annually by the Stockholm International Peace Research Institute.[20] This same data source was used to construct measure 10.2.1 of the RAND Security Cooperation Prioritization and Propensity Matching Tool. During the 2013 study, RAND researchers added a country's arms imports (denominated in millions of U.S. dollars) from Russia and China in 2016 and 2017 and divided this sum by the total value of all of the country's arms imports.[21] They then averaged the resulting percentages to provide an overall percentage of arms imports that came from Russia and China over those two years of available data.

For our effort, countries were coded as "none" if they did not receive any arms imports from Russia and China in 2016 and 2017. Countries were coded as "low" if they received less than 33 percent of the total value of their arms imports from Russia and China in 2016 and 2017.

---

[17] U.S. Bureau of Economic Analysis, "U.S. Direct Investment Abroad: Balance of Payments and Direct Investment Position Data on a Historical Cost Basis," webpage, September 1, 2018a; and U.S. Bureau of Economic Analysis, "Foreign Direct Investment in the U.S.: Balance of Payments and Direct Investment Position Data on a Historical-Cost Basis," webpage, September 1, 2018b.

[18] Gallup, *Rating World Leaders: 2018—The U.S. vs. Germany, China and Russia*, Washington, D.C., 2018.

[19] U.S. Department of State, *Voting Practices in the United Nations 2016: Report to Congress Submitted Pursuant to Public Laws 101-246 and 108-447*, Washington, D.C., August 2017; and U.S. Department of State, *Voting Practices in the United Nations 2017: Report to Congress Submitted Pursuant to Public Laws 101-246 and 108-447*, Washington, D.C., April 2018. In 2018, the State Department's Bureau of International Organization Affairs modified its methodology for assessing voting coincidence, as described in the April 2018 report.

[20] Stockholm International Peace Research Institute, undated.

[21] Paul et al., 2013.

Countries were coded as "medium" if the value was between 33 percent and 67 percent , and countries were coded as "high" if the value was more than 67 percent.[22]

Host-Nation Economic Cooperation with China

Just as arms imports are used to identify possible sources of leverage that near-peer competitors might exact over potential host nations in the secure sphere, SPRAT evaluates what share official and unofficial Chinese development assistance represents relative to a country's overall economy, as measured by aggregate GDP over the past decade. This construct draws from a data set on official and unofficial Chinese development flows released by AidData, a research lab at the College of William and Mary that synthesizes large volumes of open-source material to track more than 5,000 Chinese development projects and pledges of support to 138 countries and territories between 2000 and 2014. These figures include 4,304 Chinese development projects that were at various stages of implementation and completion by the end of 2014 and that amount to approximately $351 billion worldwide, as well as an additional 630 pledges of support amounting to an estimated $137 billion.[23]

This data set was constructed using the Tracking Underreported Financial Flows methodology developed by Strange, Cheng, et al., in 2017, which triangulates information across various English, Chinese, and local-language news reports, official statements from Chinese government agencies, aid and debt information management systems in recipient countries, and in-depth field research to identify underreported sources of Chinese official development assistance and other official flows.[24]

To approximate what share these official and unofficial aid flows represent to a host nation's overall economy, SPRAT aggregates the total value (denominated in millions of 2004 U.S. dollars) that each country received from official and unofficial development projects and pledges of support between 2000 and 2014 and divides this sum by an aggregation of each country's annual GDP (delimited in millions of 2004 U.S. dollars) as reported by the World Bank over the same time period.[25] This aggregation approach was taken to account for the fact that a Chinese development project's dollar amount was reported in a single year but may have been dispersed over multiple years in the data set.

In SPRAT, countries were coded as "none" if they did not receive any official or unofficial Chinese development assistance or pledges of support from 2000 to 2014. Countries were coded as "low" if they received official or unofficial Chinese development assistance or pledges of

---

[22] We used the 2016–2017 time frame in this analysis because the data were readily available. A more complete analysis should include a longer-term look (for example, 10–15 years) to gain a longer-term perspective.

[23] Dreher et al., 2017. These data have been used in research reports and peer-reviewed publications, including Strange, Dreher, et al., 2017.

[24] Strange, Cheng, et al., 2017.

[25] World Bank, "GDP (Current US$)," webpage, undated.

support from 2000 to 2014 in total value amounts that represented less than 1 percent of the country's GDP over the same time frame. Countries were coded as "medium" if the value represented between 1 and 3 percent, and countries were coded as "high" if the value represented more than 3 percent.

U.S. Overseas Presence

Information on the current presence of U.S. forces overseas is derived from the declassified FY 2017 Base Structure Report, as well as the number of active duty military and permanent DoD civilian personnel deployments overseas, as reported by the Defense Manpower Data Center in June 2018.[26]

In SPRAT, countries were coded as "yes" and color-coded green if they appeared in the FY 2017 Base Structure Report. According to the report's display criteria, "Sites located in a foreign country must be larger than 10 acres **OR** have a [Plant Replacement Value] greater than $10 million. Sites not meeting these criteria are aggregated as an 'Other' location within each state or country."[27] Countries containing sites that are not larger than 10 acres and do not have a plant replacement value of $10 million or more are designated with an asterisk ("yes*").

Countries were coded as "yes" and color-coded yellow if they were reported to have at least a token (non-zero) level of DoD-appropriated funding for permanently assigned military and civilian personnel. Although this presence may be insignificant when compared with that of permanent basing, the stationing of even a token U.S. presence in-country reflects a country's willingness to maintain U.S. assets on its soil. Countries that were coded as "yes**" reflect those where the presence of U.S. forces is tied to ongoing contingency operations.

Countries that did not meet any of the "yes" criteria were coded "no" for lacking any publicly reported U.S. force presence within their territory.

Access Challenges (Peacetime)

Information on a country's previous experience with hosting U.S. forces and maintaining a permanent U.S. basing presence during peacetime operations is derived from RAND researchers' comprehensive analysis of political challenges to U.S. overseas military access during peacetime and contingency operations between 1945 and 2014.[28] In the related report, Pettyjohn and Kavanagh document all instances in which a host nation restricted and evicted U.S. forces stationed overseas during peacetime.

In our SPRAT analysis, a country was coded as "none" if there were no peacetime access challenges to U.S. military presence there. It should be noted, however, that this categorization

---

[26] DoD, 2018b; Defense Manpower Data Center, undated.

[27] DoD, 2018b.

[28] Pettyjohn and Kavanagh, 2016.

includes both (1) host nations that did not impose any restrictions or evictions on the United States after permitting access and (2) countries that never offered access to begin with. Although a planner ideally would want to know which countries granted peacetime access without subsequently imposing peacetime restrictions or evictions on U.S. forces in their borders, data limitations prevent us and previous researchers from disaggregating these host nations from countries that did not receive a formal request for U.S. military access overall. Planners should therefore interpret the absence of peacetime challenges as neither a positive nor negative attribute, which is why this coding is assigned a neutral gray in our color-coding scheme.

A country was coded as "evicted" if it revoked previously granted basing rights in their entirety. This categorization does not include circumstances in which the United States decided to reduce or withdraw its overseas presence for reasons other than relations with or constraints imposed by the host nation.

A country was coded as "restricted" if it curtailed U.S. basing rights by demanding *consultation* for making changes to its military presence; altering the *duration* of the lease agreement; reasserting *sovereignty* or jurisdiction over U.S. facilities or U.S. personnel use; imposing *limits on types of forces* or activities; or imposing *contraction* that reduces the size of the U.S. military presence, in terms of the number of facilities that U.S. forces have access to or the number of platforms permitted at any one time.[29] We depart from Pettyjohn and Kavanagh's classification to include under the "restricted" banner the countries coded as demanding increased *quid pro quo* whereby the United States is compelled to provide larger aid packages or pay higher rents to maintain access. Although this is not a restriction on access in the formal sense, it does depict a host nation threatening restrictions or even evictions and thus represents a curtailment of the country's willingness to host U.S. forces.

The restrictions that the host nation imposed are relative to the terms of the original access agreement establishing the terms of U.S. military presence. The imposition of restrictions, therefore, should not be interpreted as an indication of the substantive size and scale of U.S. presence in the country. It could be the case that a host nation, such as the United Kingdom, that imposes restrictions on an agreement that has historically given relatively wide latitude to a large U.S. force presence still offers the United States greater flexibility than U.S. forces in host nations that have not mandated such restrictions but have imposed more-stringent terms under the original baseline access agreement.

When a country's coding is accompanied by a single asterisk (*), it indicates that the latest peacetime access challenges took place prior to the collapse of the Soviet Union (December 1991). Such instances are more historically confined to broader Cold War dynamics of the superpower competition and a period when the United States had different basing posture. More-recent history, such as the resumption of relations with Vietnam (1997), is more likely to influence contemporary events.

---

[29] Pettyjohn and Kavanagh, 2016, pp. 57–58.

At different times since World War II, countries have imposed different kinds of restrictions and have, on occasion, evicted U.S. forces after imposing a series of restrictions. For example, Libya repeatedly demanded increased foreign aid or rent payments (1954, 1960, 1964, 1967) before evicting U.S. forces in 1969.[30] Similarly, Panama imposed restrictions on U.S. basing access before evicting U.S. forces in 1999. The existing coding reflects the highest degree of challenge historically imposed on U.S. basing presence.

There are other instances in which the U.S. military is evicted and has subsequently had its basing rights restored (as was the case with the Philippines in 2014). It is therefore possible that a country where the United States currently maintains an active, permanent overseas basing presence may still be classified for this category as having been "evicted" if this had happened historically in the post–Cold War era. It may appear misleading or contradictory to retain this categorization if basing rights have subsequently been restored. But the RAND researchers concluded that it is valuable for the planner to have a sense of long-range changes in the basing relationship in order to better anticipate potential fault lines over the long term.

In the case of Turkey and Thailand (both designated with a double asterisk [**]), both countries evicted the United States from permanent basing presence during the Cold War and subsequently restored U.S. basing rights on which they imposed restrictions during the post–Cold War era. Both Cold War evictions and post–Cold War restrictions are represented in the coding for these two countries.

Access Challenges (Contingency)

Information on access challenges during contingency operations similarly draws on the Pettyjohn and Kavanagh analysis.[31] Given that the authors identify meaningful differences in the types and sources of access restrictions and denials across peacetime and contingency contexts, we felt it was important to present these data separately. As Pettyjohn and Kavanagh note, this information "includes only documented non-routine requests to access another country's territory or airspace for a particular operation. . . . The data set, therefore, excludes steady-state requests for access to support forward-based forces and regular peacetime exercises that were not in response to a particular stimulus."[32]

A country was coded as "denied" if a formal U.S. request for access during a contingency operation was denied in its entirety by a potential host nation between 1945 and 2014. A country was coded as "restricted" if a formal request was authorized with less permissive or conditional

---

[30] Pettyjohn and Kavanagh, 2016, Appendix A.

[31] Pettyjohn and Kavanagh, 2016, p. 139.

[32] Pettyjohn and Kavanagh, 2016, p. 69.

forms of access that impose "meaningful limits on the type, size, location, or operation of U.S. forces."[33]

A country was coded as "none" if it did not create access challenges for the U.S. military during contingency operations. However, this categorization includes both (1) host nations that did not impose any restrictions or denials on the United States after permitting unrestricted access and (2) countries that never offered access to begin with. Although a planner ideally would want to know which countries granted unrestricted access without subsequently imposing restrictions or denials on U.S. forces, data limitations prevent us and previous researchers from disaggregating host nations that granted unrestricted access from countries that never received a formal U.S. access request. Planners should therefore interpret the absence of challenges during contingency operations as neither a positive nor negative attribute, which is why this coding is assigned a neutral gray in our color-coding scheme.

---

[33] Pettyjohn and Kavanagh, 2016, pp. 192–193.

References

Amouzegar, Mahyar A., Ronald G. McGarvey, Robert S. Tripp, Louis Luangkesorn, Thomas Lang, and Charles Robert Roll, Jr., *Evaluation of Options for Overseas Combat Support Basing*, Santa Monica, Calif.: RAND Corporation, MG-421-AF, 2006. As of December 10, 2019:
https://www.rand.org/pubs/monographs/MG421.html

Amouzegar, Mahyar A., Robert S. Tripp, Ronald G. McGarvey, Edward W. Chan, and C. Robert Roll, Jr., *Supporting Air and Space Expeditionary Forces: Analysis of Combat Support Basing Options*, Santa Monica, Calif.: RAND Corporation, MG-261-AF, 2004. As of December 10, 2019:
https://www.rand.org/pubs/monographs/MG261.html

Arvis, Jean-Francois, Lauri Ojala, Christina Wiederer, Ben Shepherd, Anasuya Raj, Karlygash Bairabayeva, and Toumas Kiiski, *Connecting to Compete 2018: Trade Logistics in the Global Economy*, Washington, D.C.: International Bank for Reconstruction and Development/World Bank, 2018.

Bandaly, Dia, Ahmet Satir, Yasemin Kahyaoglu, and Latha Shanker, "Supply Chain Risk Management – I: Conceptualization, Framework and Planning Process," *Risk Management*, Vol. 14. No. 4, November 2012, pp. 249–271.

Bartlett, Christopher A., and Sumantra Ghosbal, "Managing Across Borders: New Strategic Requirements," *Sloan Management Review*, Vol. 28, No. 4, Summer 1987, pp. 7–17.

Bhatnagar, Rohit, Jayanth Jayaram, and Yue Cheng Phua, "Relative Importance of Plant Location Factors: A Cross National Comparison Between Singapore and Malaysia," *Journal of Business Logistics*, Vol. 24, No. 1, 2003, pp. 147–170.

Blanchard, Ben, "China Formally Opens First Overseas Military Base in Djibouti," Reuters, August 1, 2017.

Bolsworth, James, "Cold War Defense Treaty Under Fire in Latin America," *Christian Science Monitor*, June 8, 2012.

Carney, Steven A., *Allied Participation in Operation Iraqi Freedom*, Washington, D.C.: Center for Military History, 2011.

Carr, Rick, and Sam Pearson, "Unconventional Drilling Requires Managing Transportation Logistics," *Oil and Gas Journal*, June 4, 2012.

Chima, Christopher M., "Supply-Chain Management Issues in the Oil and Gas Industry," *Journal of Business & Economics Research*, Vol. 5, No. 6, June 2007.

Daalder, Ivo H., and James G. Stavridis, "NATO's Victory in Libya: The Right Way to Run an Intervention," *Foreign Affairs*, Vol. 91, No. 2, March/April 2012, pp. 2–7.

Daly, E., and T. Fore, *Operationalizing Army Pre-Positioned Stocks (APS)*, Rock Island Arsenal, Ill.: U.S. Army Sustainment Command, March 30, 2017, Not available to the general public.

Defense Manpower Data Center, "DoD Personnel, Workforce Reports & Publications," webpage, undated. As of March 20, 2020:
https://www.dmdc.osd.mil/appj/dwp/dwp_reports.jsp

Department of the Air Force, *Air Force War Reserve Materiel (WRM) Policies and Guidance*, Air Force Instruction 25-101, January 14, 2015.

———, *Air Force Materiel Management*, Air Force Instruction 23-101, December 12, 2016.

Department of the Navy, "Budget Materials," webpage, undated. As of September 1, 2018:
http://www.secnav.navy.mil/fmc/fmb/Pages/Fiscal-Year-2018.aspx

DoD—*See* U.S. Department of Defense.

Dreher, Axel, Andreas Fuchs, Bradley Parks, Austin M. Strange, and Michael J. Tierney, *Aid, China, and Growth: Evidence from a New Global Development Finance Dataset*, Williamsburg, Va.: AidData, Working Paper 46, October 2017.

Federal Emergency Management Agency, *2018–2022 Strategic Plan*, Washington, D.C., March 15, 2018.

FEMA—*See* Federal Emergency Management Agency.

Foss, Nicolai J., and Peter G. Klein, "Why Managers Still Matter," *Sloan Management Review*, Vol. 56, No. 1, Fall 2014, pp. 73–80.

Freiden, Jeffrey A., *Global Capitalism: Its Fall and Rise in the Twentieth Century*, New York: W.W. Norton & Company, 2006.

Freidman, Thomas, *The World Is Flat: A Brief History of the Twenty-First Century*, New York: Farrar, Straus, and Giroux, 2005.

Gallup, *Rating World Leaders: 2018—The U.S. vs. Germany, China and Russia*, Washington, D.C., 2018.

Georlett, Jacqueline, and Bruce Daasch, "Army Pre-Positioned Stocks Support Army Readiness," *Army Sustainment*, May–June 2017.

Global Coalition Against Daesh, "Partners," webpage, undated. As of September 13, 2018:
http://theglobalcoalition.org/en/partners

Hamel, Gary, and C. K. Prahalad, "Do You Really Have a Global Strategy?" *Harvard Business Review*, July 1985, pp. 139–148.

Harlan, Michael J., *Force Projection Logistics Atrophy: Affliction and Treatment*, Carlisle Barracks, Pa.: U.S. Army War College, March 2011.

Headquarters, Department of the Army, *Army Pre-Positioned Operations*, Army Techniques Publication 3-35.1, Washington, D.C., October 2015.

Hout, Thomas, Michael E. Porter, and Eileen Rudden, "How Global Companies Win Out," *Harvard Business Review*, September 1982, pp. 98–108.

HQDA—*See* Headquarters, Department of the Army.

Jane's Military and Security Assessments Intelligence Centre, "Global Country Risk Ratings," IHS Markit, August 14, 2018.

Joint Chiefs of Staff, *Logistics Planning Guidance for Pre-Positioned War Reserve Materiel*, Chairman of the Joint Chiefs of Staff Instruction 4310.01D, Washington, D.C., December 30, 2016.

Kaly, Ursula, Craig Pratt, and Jonathan Mitchell, *The Demonstration Environmental Vulnerability Index (EVI) 2004*, Suva, Fiji: South Pacific Applied Geoscience Commission, Technical Report 384, 2004.

Karimi, Jahangir, and Benn R. Konsynski, "Globalization and Information Management Strategies," *Journal of Management Information Systems*, Vol. 7, No. 4, Spring 1991, pp. 7–26.

Kavanagh, Jennifer, *U.S. Security Agreements in Force Since 1955: Introducing a New Database*, Santa Monica, Calif.: RAND Corporation, RR-736-AF, 2014. As of December 10, 2019:
https://www.rand.org/pubs/research_reports/RR736.html

Kemp, John, "U.S. Drilling Costs Start to Rise as Rig Count Climbs: Kemp," Reuters, July 14, 2017.

Kleindorfer, Paul R., and Germaine H. Saad, "Managing Disruption Risks in Supply Chains," *Production and Operations Management*, Vol. 14, No. 1, Spring 2005, pp. 53–68.

Lang, Thomas, *Defining and Evaluating Reliable Options for Overseas Combat Support Basing*, Santa Monica, Calif.: RAND Corporation, RGSD-250, 2009. As of December 10, 2019:
https://www.rand.org/pubs/rgs_dissertations/RGSD250.html

Lang, Thomas E., and Ronald G. McGarvey, "Determining Reliable Networks of Prepositioning Materiel Warehouses for Public-Sector Rapid Response Supplies," *Advances in Operations Research*, 2016, pp. 1–20.

Lee, Hau L., and Corey Billington, "Managing Supply Chain Inventory: Pitfalls and Opportunities," *Sloan Management Review*, Vol. 33, No. 3, Spring 1992, pp. 65–73.

Lee, Hau L., V. Padmanabhan, and Seugjin Whang, "Information Distortion in Supply Chains: The Bullwhip Effect," *Management Science*, Vol. 42, No. 4, 1997, pp. 546–558.

Loredo, Elvira N., Ryan Schwankhart, Abby Doll, Karlyn D. Stanley, Marta Kepe, Bryan Boling, Luke Muggy, Endy M. Daehner, and Simon Véronneau, *Challenges to Timely Movement of Forces in Europe: Moving Heavy Equipment, Ammunition, and Fuel*, Santa Monica, Calif.: RAND Corporation, 2018, Not available to the general public.

Lostumbo, Michael J., Michael J. McNerney, Eric Peltz, Derek Eaton, David R. Frelinger, Victoria A. Greenfield, John Halliday, Patrick Mills, Bruce R. Nardulli, Stacie L. Pettyjohn, Jerry M. Sollinger, and Stephen M. Worman, *Overseas Basing of U.S. Military Forces: An Assessment of Relative Costs and Strategic Benefits*, Santa Monica, Calif.: RAND Corporation, RR-201-OSD, 2013. As of December 10, 2019: https://www.rand.org/pubs/research_reports/RR201.html

Lynch, Kristin F., John G. Drew, Robert S. Tripp, and C. Robert Roll, Jr., *Supporting Air and Space Expeditionary Operations: Lessons from Operation Iraqi Freedom*, Santa Monica, Calif.: RAND Corporation, MG-193-AF, 2005. As of December 10, 2019: https://www.rand.org/pubs/monographs/MG193.html

MacCormack, Alan David, Lawrence James Newman III, and Donald B. Rosenfield, "The New Dynamics of Global Manufacturing Site Location," *Sloan Management Review*, Vol. 35, No. 4, Summer 1994, pp. 69–80.

Malone, Thomas W., "Making the Decision to Decentralize," Harvard Business School Working Knowledge, March 29, 2004.

Manuj, Ila, and John T. Mentzer, "Global Supply Chain Risk Management," *Journal of Business Logistics*, Vol. 29, No. 1, 2008, pp. 135–153.

"Markets: Oil and Gas Drilling," *New York Times*, undated. As of December 16, 2019: https://markets.on.nytimes.com/research/markets/usmarkets/industry.asp?industry=50131

McGarvey, Ronald G., Robert S. Tripp, Rachel Rue, Thomas Lang, Jerry M. Sollinger, Whitney A. Conner, and Louis Luangkesorn, *Global Combat Support Basing: Robust Prepositioning Strategies for Air Force War Reserve Materiel*, Santa Monica, Calif.: RAND Corporation, MG-902-AF, 2010. As of December 10, 2019: https://www.rand.org/pubs/monographs/MG902.html

Mills, Patrick, John G. Drew, John A. Ausink, Daniel M. Romano, and Rachel Costello, *Balancing Agile Combat Support Manpower to Better Meet the Future Security Environment*,

Santa Monica, Calif.: RAND Corporation, RR-337-AF, 2014. As of December 10, 2019:
https://www.rand.org/pubs/research_reports/RR337.html

Mills, Patrick, James A. Leftwich, Kristin Van Abel, and Jason Mastbaum, *Estimating Air Force Deployment Requirements for Lean Force Packages: A Methodology and Decision Support Tool Prototype*, Santa Monica, Calif.: RAND Corporation, RR-1855-AF, 2017. As of December 10, 2019:
https://www.rand.org/pubs/research_reports/RR1855.html

Mshelia, James B., and John R. Anchor, "Political Risk Assessment by Multinational Corporations in African Markets: A Nigerian Perspective," *Thunderbird International Business Review*, February 2018, pp. 1–10.

Multilateral Investment Guarantee Agency, *World Investment and Political Risk 2013*, Washington, D.C.: International Bank for Reconstruction and Development/World Bank, December 2013.

North Atlantic Treaty Organization, "Archive ISAF Placemats: NATO and Afghanistan," webpage, May 23, 2017a. As of December 10, 2019:
https://www.nato.int/cps/en/natohq/107995.htm

———, "Montenegro Joins NATO as 29th Ally," press release, June 9, 2017b.

Office of the Chief of Naval Operations, *Navy War Reserve Materiel Program*, Office of the Chief of Naval Operations Instruction 4080.11D, January 21, 1999.

Paul, Christopher, Michael Nixon, Heather Peterson, Beth Grill, and Jessica Yeats, *The RAND Security Cooperation Prioritization and Propensity Matching Tool*, Santa Monica, Calif.: RAND Corporation, TL-112-OSD, 2013. As of December 10, 2019:
https://www.rand.org/pubs/tools/TL112.html

Peltz, Eric, Marc L. Robbins, Kenneth J. Girardini, Rick Eden, John M. Halliday, and Jeffrey Angers, *Sustainment of Army Forces in Operation Iraqi Freedom: Major Findings and Recommendations*, Santa Monica, Calif.: RAND Corporation, MG-342-A, 2005. As of December 10, 2019:
https://www.rand.org/pubs/monographs/MG342.html

Pettyjohn, Stacie L., and Jennifer Kavanagh, *Access Granted: Political Challenges to the U.S. Overseas Military Presence, 1945–2014*, Santa Monica, Calif.: RAND Corporation, RR-1339-AF, 2016. As of December 10, 2019:
https://www.rand.org/pubs/research_reports/RR1339.html

Pettyjohn, Stacie L., and Alan J. Vick, *The Posture Triangle: A New Framework for U.S. Air Force Global Presence*, Santa Monica, Calif.: RAND Corporation, RR-402-AF, 2013. As of

December 10, 2019:
https://www.rand.org/pubs/research_reports/RR402.html

Porter, Michael E., "Competition in Global Industries: A Conceptual Framework," in Michael E. Porter, ed., *Competition in Global Industries*, Boston, Mass.: Harvard Business School Press, 1986a, pp. 15–60.

———, "Introduction and Summary," in Michael E. Porter, ed., *Competition in Global Industries*, Boston, Mass.: Harvard Business School Press, 1986b, pp. 1–11.

Richardson, Deila A., Sander de Leeuw, and Wout Dullaert, "Factors Affecting Global Inventory Prepositioning Locations in Humanitarian Operations—A Delphi Study," *Journal of Business Logistics*, Vol. 37, No. 1, 2016, pp. 59–74.

Rodrick, Dani, *The Globalization Paradox: Democracy and the Future of the World Economy*, New York: W.W. Norton & Company, 2011.

Rosello, Anthony D., Muharrem Mane, Jeffrey R. Brown, Emmi Yonekura, Henry Hargrove, Alexander Halman, Paul W. Mayberry, Dan Madden, Nahom M. Beyene, and C. R. Anderegg, *Air Force Readiness Reporting Relation to Operations Tempo and Suggested Improvements*, Santa Monica, Calif.: RAND Corporation, 2018, Not available to the general public.

Rugman, Alan M., and Alain Verbeke, "A Perspective on Regional and Global Strategies of Multinational Enterprises," *Journal of International Business Studies*, Vol. 35, No. 1, January 2004, pp. 3–18.

Scott, Patrick, and Boxi Xu, *Multi-Echelon Inventory Modeling and Supply Redesign*, Cambridge, Mass.: Massachusetts Institute of Technology, 2017.

Singh, Gurdeep, Anant Tripathi, Anupriya Srivastava, and Mahesh Iyer, "Integrated Supply Chain Outsourcing: Expanding the Role of Third Party Logistics in the Upstream Industry," Society of Petroleum Engineers Oil & Gas India Conference and Exhibition, Mumbai, November 24–26, 2015.

Snyder, Don, and Patrick Mills, *Supporting Air and Space Expeditionary Forces: A Methodology for Determining Air Force Deployment Requirements*, Santa Monica, Calif.: RAND Corporation, MG-176-AF, 2004. As of December 10, 2019:
https://www.rand.org/pubs/monographs/MG176.html

Steinle, Claus, and Holger Schiele, "Limits to Global Sourcing? Strategic Consequences of Dependency on International Suppliers: Cluster Theory, Resource-Based View and Case Studies," *Journal of Purchasing and Supply Management*, Vol. 14, No. 1, March 2018, pp. 3–14.

Stieglitz, Joseph, *Globalization and Its Discontents*, New York: W.W. Norton & Company, 2002.

Stockholm International Peace Research Institute, "SIPRI Arms Transfers Database," webpage, undated. As of December 10, 2019:
https://www.sipri.org/databases/armstransfers

Strange, Austin M., Alex Dreher, Andreas Fuchs, Bradley Parks, and Michael Tierney, "Tracking Underreported Financial Flows: China's Development Assistance and the Aid-Conflict Nexus Revisited," *Journal of Conflict Resolution*, Vol. 61, No. 5, 2017, pp. 935–963.

Strange, Austin M., Mengfan Cheng, Brooke Russell, Siddhartha Ghose, and Bradley Parks, *AidData Methodology: Tracking Underreported Financial Flows (TUFF)*, Version 1.3, Williamsburg, Va.: AidData, October 2017.

Strohecker, Karin, "Sao Tome Signs Memorandum with China on Deep Sea Port," Reuters UK, October 12, 2015.

Thaler, David E., Beth Grill, Jefferson P. Marquis, Jennifer D. P. Moroney, Heather Peterson, Lisa Saum-Manning, and Ilana Blum, *Assessing Partner-Nation Air, Space, and Cyber Capabilities: Supporting Development of Security Cooperation Strategy for the U.S. Air Force Flight Plan*, Santa Monica, Calif.: RAND Corporation, 2018, Not available to the general public.

Thomas, Brent, Mahyar A. Amouzegar, Rachel Costello, Robert A. Guffey, Andrew Karode, Christopher Lynch, Kristin F. Lynch, Ken Munson, Chad J. R. Ohlandt, Daniel M. Romano, Ricardo Sanchez, Robert S. Tripp, and Joseph V. Vesely, *Project AIR FORCE Modeling Capabilities for Support of Combat Operations in Denied Environments*, Santa Monica, Calif.: RAND Corporation, RR-427-AF, 2015. As of November 30, 2020:
https://www.rand.org/pubs/research_reports/RR427.html

Thomas, Brent, Bradley DeBlois, Katherine C. Hastings, Beth Grill, Anthony DeCicco, Sarah A. Nowak, and John A. Hamm, *Developing a Global Posture for Air Force Expeditionary Medical Support*, Santa Monica, Calif.: RAND Corporation, forthcoming, Not available to the general public.

U.S. Bureau of Economic Analysis, "U.S. Direct Investment Abroad: Balance of Payments and Direct Investment Position Data on a Historical Cost Basis," webpage, September 1, 2018a. As of September 13, 2018:
https://www.bea.gov/international/di1usdbal

———, "Foreign Direct Investment in the U.S.: Balance of Payments and Direct Investment Position Data on a Historical-Cost Basis," webpage, September 1, 2018b. As of September

13, 2018:
https://www.bea.gov/international/di1fdibal

U.S. Census Bureau, "U.S. Trade in Goods by Country," webpage, undated. As of September 13, 2018:
https://www.census.gov/foreign-trade/balance/index.html

U.S. Department of Defense, *Summary of the 2018 National Defense Strategy of the United States of America: Sharpening the American Military's Competitive Edge*, Washington, D.C., 2018a.

———, *Base Structure Report—Fiscal Year 2017 Baseline: A Summary of the Real Property Inventory*, Washington, D.C., April 4, 2018b.

U.S. Department of Homeland Security, *National Response Framework*, 2nd ed., Washington, D.C., May 2013, Figure 5.

U.S. Department of State, *Voting Practices in the United Nations 2016: Report to Congress Submitted Pursuant to Public Laws 101-246 and 108-447*, Washington, D.C., August 2017.

———, *Voting Practices in the United Nations 2017: Report to Congress Submitted Pursuant to Public Laws 101-246 and 108-447*, Washington, D.C., April 2018.

Vantrappen, Herman, and Frederic Wirtz, "When to Decentralize Decision Making, and When Not To," *Harvard Business Review*, December 26, 2017.

Verbeke, Alain, and Christian Giesler Asmussen, "Global Local, or Regional? The Locus of MNE Strategies," *Journal of Management Studies*, Vol. 53, No. 6, September 2016, pp. 1051–1075.

Vick, Alan, David Orletsky, Bruce Pirnie, and Seth Jones, *The Stryker Brigade Combat Team: Rethinking Strategic Responsiveness and Assessing Deployment Options*, Santa Monica, Calif.: RAND Corporation, MR-1606-AF, 2002. As of December 10, 2019:
https://www.rand.org/pubs/monograph_reports/MR1606.html

Watt, Louise, "China Resumes Ties with Sao Tome in Triumph Over Taiwan," Associated Press, December 26, 2016.

Wogan, David, "Oil and Natural Gas Drilling Rigs Are Moving In at a Furious Pace," *Scientific American*, March 5, 2014.

World Bank, "GDP (Current US$)," webpage, undated. As of September 13, 2018:
https://data.worldbank.org/indicator/NY.GDP.MKTP.CD

———, Logistics Performance Index, Washington, D.C., 2018.

Yip, George S., "Global Strategy . . . In a World of Nations?" *Sloan Management Review*, Vol. 31, No. 1, Fall 1989, pp. 29–41.

————, *Total Global Strategy II: Updated for the Internet and Service Era*, Upper Saddle River, N.J.: Prentice Hall, 2003.

Zakaria, Fareed, *The Post-America World*, New York: W.W. Norton & Company, 2008.